Portland cement

A modern dry process plant with precalciner (courtesy of Krupp Polysius)

Portland cement

Composition, production and properties

2nd edition

G.C. Bye

Thomas Telford

Published by Thomas Telford Publishing, Thomas Telford Limited, 1 Heron Quay, London E14 4JD.

URL: http://www.t-telford.co.uk

Distributors for Thomas Telford books are
USA: ASCE Press, 1801 Alexander Bell Drive, Reston, VA 20191-4400
Japan: Maruzen Co. Ltd, Book Department, 3–10 Nihonbashi 2-chome, Chuo-ku, Tokyo 103
Australia: DA Books and Journals, 648 Whitehorse Road, Mitcham 3132, Victoria

First published 1983 by Pergamon Press Ltd
Thomas Telford edition 1999

Cover shows a kiln burner's view of the flame in a cement kiln (courtesy of F.L. Smidth–Fuller)

A catalogue record for this book is available from the British Library

ISBN: 0 7277 2766 4

Typeset by MHL Typesetting Ltd, Coventry.
Printed and bound in Great Britain by Bookcraft, Midsomer Norton, Somerset.

Preface

This book is intended for use in ceramics and material science undergraduate and postgraduate courses. The reader should have had a grounding in chemistry and physics, ideally with an emphasis on phase equilibria, the kinetics of heterogeneous processes, the solid state and some aspects of surface science, such as would be provided in these courses. It is also hoped that a substantial part of the text will be of value to graduates in pure science and engineering, who may be starting related postgraduate research or entering the cement and concrete industries.

This edition has the same two objectives as the previous one. The first is to provide a basic vocabulary of Portland cement science and of its manufacture. Since any account of the properties of Portland cement is concerned mainly with its hydration, the second objective is to provide an introduction to current knowledge of hydration reactions and of physical property-microstructure relationships in hardened cement paste. In view of the extensive literature on these topics only a selection of published work has been mentioned, chosen where possible to illustrate the application of physico-chemical principles and techniques. The text has been updated and enlarged by the addition of a separate chapter dealing with cements related to Portland cement, including its blends with reactive minerals.

Authors of work quoted are acknowledged but only a selection of references are given. For those wishing to specialise, the further reading list is intended to provide an entry into the considerable literature of the subject. Some reference to original papers is included to supplement review articles, where they could be used for tutorial work or provide a source of additional electron micrographs.

The writing of this second edition was suggested by Professor F.P. Glasser to whom the author is indebted for an introduction to Thomas Telford Publishing, and for support and advice throughout. Of great value to the project were contacts with former colleagues in Blue Circle Cement, Technical Centre, generously authorised by Mr I.S.V. McKenzie and use of the library of the British Cement Association, with the kind permission of Mr J. Stevenson. I would also like to thank the Association's Librarian, Mr E. Trout and his staff for their help.

Completion of the book would not have been possible without the kindness of a significant number of people who so willingly provided suggestions and comments on parts of draft text, although, of course, any errors in using such comment are mine. Thanks are due to the following. Mr W.A. Gutteridge, Mr D.W. Hobbs and Mr M.G. Taylor (all of British Cement Association); Mr A.D.R. Brown (who also supplied a copy of a paper on sulfoaluminate cements given to a meeting of the Society of Chemical Industry in 1998), Mr R.S. Gollop and Dr G.K. Moir (all of Blue Circle Technical Centre); Mr N. Hepher (F.L. Smidth-Fuller); Professor J.H. Sharp (University of Sheffield); Dr K.L. Scrivener (Lafarge); Professor H.M. Jennings (Northwestern University, USA); and for especially sustained support Mr C.P. Kerton and Mr J.S. Lumley (both of Blue Circle Technical Centre). For valuable and detailed comment on a substantial portion of the final draft, I am particularly indebted to Professor H.F.W. Taylor. Finally, I am most grateful to Mr Alex Lazarou, Mr Paul Clements and colleagues of Thomas Telford Publishing for their patient help in bringing this project to a conclusion.

Permissions

The author is grateful for the kind donation of photographs, the donors of which are acknowledged in the captions. Particular thanks are due to Mr R. Heaney and F.L. Smidth-Fuller for the loan of an archive colour negative used for the cover of this book. Every reasonable effort has been made to obtain all copyright permissions.

For permission to reproduce figures and/or data the author thanks the following (authors are acknowledged in the captions).

Fig. 3.1 Blue Circle Industries

Fig. 3.6 The Fuller Co. (formerly licensee of the IHI-SF Process)

Fig. 3.9 Seidel/Huckauf/Stark. Technologie der Bindebaustoffe, Band 3, 1. Auflage. HUSS-MEDIEN GmbH/Verlag Bauwesen, Berlin, 1977.

Figs 3.10, 5.4, 5.6 ZKG International, Bauverlag, Wiesbaden

Figs 5.2, 5.3 The Institute of Materials, London

Figs 5.5, 8.12, Table 8.1 The American Ceramic Society, Westerville, Ohio

Fig. 6.1 The British Standards Institution. (Extracts from BS EN 196-3 reproduced under licence number PD/1999 0300. Complete copies of the standard can be obtained from BSI Customer Services, 389 Chiswick High Road, London W4 4AL.)

Figs 7.2, 8.4 Academic Press, London. Reprinted from *The Chemistry of Cements*, H.F.W. Taylor (ed.), 1964

Figs 7.3, 7.7, 8.7, 8.8, 8.17 Elsevier Science, Oxford. Reprinted from *Cement & Concrete Research*: L.J. Parrott, M. Geiker, W.A. Gutteridge and D. Killoh, Monitoring Portland cement hydration: comparison of methods. **20**, (1990), 919; L.S. Dent Glasser, E.E. Lachowski, K. Mohan and H.F.W. Taylor, A multi-method study of C_3S hydration. **8**, (1978), 733; S. Diamond, A critical comparison of mercury porosimetry and capillary condensation pore size distributions in Portland cement pastes. **1**, (1971), 531; R.F. Feldman and J.J. Beaudoin, Microstructure and strength of hydrated cements. **6**, (1976), 389; H.F.W. Taylor, Discussion of the paper 'Microstructure and strength of hydrated cements' by R.F. Feldman and J.J. Beaudoin. **7**, (1977), 465.

Figs 7.5, 9.6, 10.1 Arnold-Hodder Headline Group

Fig. 7.6   ATILH, Paris. Reprinted from Fig. 29 in Electromechanical study of the aqueous phase in C_3S hydration by A. Zelwer, in *Rev. Materiaux de Construction et Travaux Publics*, 1973, (681), 20.

Figs 7.9, 9.3   British Cement Association, Crowthorne, Berks.

Figs 8.6, 8.16   Editions Eyrolles, Paris

Fig. 8.9   *Magazine of Concrete Research*, Thomas Telford, London

Fig. 8.10   Transportation Research Board, National Research Council, Washington DC. Reprinted from R.A. Helmuth and D.H. Turk, Elastic moduli of hardened Portland cement and tricalcium silicate pastes, in Special Report 90: *Symposium on Structure of Portland Cement Paste and Concrete*, Washington DC, 1966.

Fig. 8.13 (data)   Portland Cement Association, Skokie, Illinois

Fig. 9.6   Organizing Commission, 8th International Congress on the Chemistry of Cement, Rio de Janeiro, 1986

Fig. 9.7   National Council for Cement and Building Materials, New Dehli (9th International Congress, on the Chemistry of Cement, 1993, **5**, 3)

Units

SI units are widely employed in the cement industry. The closely related Angstrom unit ($= 10^{-10}$ m) is still sometimes used for atomic sizes and crystal lattice spacings.

All temperatures quoted in this book are in degrees Celsius (centigrade) except where temperatures (T) on the thermodynamic (absolute) scale in Kelvin (K) are specified.

The gas constant, $R = 8.314$ J/K mol ($\equiv 8.314$ J K^{-1} mol^{-1}).

Energy consumption in grinding is usually expressed in kilowatt hour/tonne and 1 kWh $= 3.6 \times 10^6$ joule. In American mineral processing literature, the short ton (= 2000 lb or 907.125 kg) is frequently used.

Strengths are given here in N/mm^2, but the megapascal (MPa) is also employed:

$$1 \text{ MPa} = 1 \text{ MN/m}^2 = 1 \text{ N/mm}^2$$

Other common pressure units are:

1 bar $= 10^5$ N/m^2
1 torr $= 133.322$ N/m^2 (1 atm/760)

Magnetic flux density: 1 Tesla $= 10^4$ Gauss
All percentages are mass/mass except where otherwise stated.

Contents

1. Introduction and composition of Portland cement   1
    1.1   Introduction, 1
            1.1.1   Cementitious systems, 1
            1.1.2   Portland cement, 2
            1.1.3   Historical aspects, 3
    1.2   Composition of Portland cement, 5
            1.2.1   Nomenclature, 5
            1.2.2   High-temperature phase equilibria, 6
            1.2.3   Compound composition of Portland cement, 8
    1.3   Polymorphism and solid solution in clinker compounds, 9
            1.3.1   Tricalcium silicate (alite), 11
            1.3.2   Dicalcium silicate (belite), 11
            1.3.3   Tricalcium aluminate, 13
            1.3.4   Calcium aluminoferrite: the ferrite phase, 13
    1.4   Minor clinker compounds, 14

2. Raw materials   15
    2.1   Raw materials, 15
            2.1.1   Calcareous component, 15
            2.1.2   Argillaceous component, 17
            2.1.3   Multicomponent mixes, 19
            2.1.4   Gypsum, 19
            2.1.5   Coal ash, 19
    2.2   Proportioning of raw materials, 20
            2.2.1   Calculation of the composition of a kiln feed, 20
    2.3   Reactivity of the raw materials, 23
            2.3.1   Model for combinability, 26
    2.4   Physical properties of raw materials, 28
            2.4.1   Grindability, 28

3. **Production of cement clinker** **31**
3.1 Introduction, 31
3.2 Preparation of kiln feed, 31
    3.2.1 Wet and semi-wet processes, 32
    3.2.2 Dry and semi-dry processes, 34
3.3 Pyroprocessing: principal manufacturing processes, 36
    3.3.1 Introduction, 36
    3.3.2 Wet and semi-wet processes, 37
    3.3.3 Dry processes, 38
    3.3.4 Semi-dry (Lepol) process, 39
    3.3.5 Clinker cooling, 40
    3.3.6 Refractories, 40
3.4 Pyroprocessing: physical and chemical processes involved, 41
    3.4.1 Preheating, 41
    3.4.2 Calcining, 42
    3.4.3 Clinkering (sintering in the presence of a liquid phase), 46
    3.4.4 Cooling, 49
3.5 Thermal efficiency of pyroprocessing, 49
    3.5.1 Process control, 49
    3.5.2 The heat balance — process efficiency, 51
    3.5.3 Electric power consumption, 52

4. **Characterisation of Portland cement clinker** **53**
4.1 Introduction, 53
4.2 Chemical analysis by selective dissolution, 54
4.3 Optical microscopy, 54
    4.3.1 Characteristics of the principal clinker phases, 55
        4.3.1.1 Alite (C_3S — density $3150\,kg/m^3$), 55
        4.3.1.2 Belite (C_2S — density $3280\,kg/m^3$), 57
        4.3.1.3 Interstitial phases, 57
        4.3.1.4 Minor phases, 58
    4.3.2 Quantitative determination of phase composition, 58
4.4 X-ray diffraction, 58
    4.4.1 Quantitative X-ray diffraction analysis (QXDA), 60
4.5 Electron microscopy, 62
    4.5.1 Backscattered electron (BSE) imaging, 63
    4.5.2 X-ray microanalysis, 63
4.6 Concluding remarks, 64

5.   Grinding and fineness of cement   67
    5.1   Cement milling, 67
          5.1.1   Factors influencing the grindability of clinker, 71
          5.1.2   Minor additional constituents, 72
          5.1.3   Addition of gypsum, 72
    5.2   Fineness of cement, 73
          5.2.1   Determination of surface area, 74
          5.2.2   Particle size distribution, 76

6.   Tests of cement quality   79
    6.1   Introduction, 79
    6.2   Chemical composition, 81
    6.3   Setting times, 81
    6.4   Compressive strength, 82
    6.5   Workability, 85
    6.6   Soundness, 87
    6.7   Heat of hydration, 89
    6.8   Concluding remarks — durability of concrete, 92

7.   The hydration of Portland cement   95
    7.1   Introduction, 95
          7.1.1   General introduction, 95
          7.1.2   Methods of investigating the kinetics of cement hydration, 97
    7.2   Hydration of the individual phases in Portland cement, 103
          7.2.1   Tricalcium silicate, 103
                  7.2.1.1   Kinetics of tricalcium silicate hydration, 109
                  7.2.1.2   Origin of the dormant (induction) period, 112
          7.2.2   Dicalcium silicate, 115
          7.2.3   Tricalcium aluminate, 116
          7.2.4   Calcium aluminoferrite, 119
    7.3   Hydration of Portland cement, 120
    7.4   Hydration at elevated temperatures, 125
          7.4.1   Delayed (secondary) ettringite formation, 126
    7.5   Concluding remarks, 128

8.   The nature of hardened cement paste   131
    8.1   Microstructure of hardened cement paste (hcp), 131
    8.2   Surface area and porosity of hardened cement paste, 136

8.2.1   Surface area, 136
8.2.2   Porosity, 139
8.2.3   Pore size distribution, 141
8.3   Physical properties of hardened cement paste, 144
8.3.1   Permeability, 144
8.3.2   Compressive strength, 146
8.3.3   Elastic and inelastic properties, 149
8.3.4   Drying shrinkage, 153
8.4   Nature of the solid–solid bond in hardened cement paste, 158
8.5   Modelling microstructure–physical property development in hydrating hcp, 160
8.6   Concluding remarks, 161

9.   Portland cements and related blended cements   163
9.1   Introduction, 163
9.2   Type I and related Portland cements, 163
9.2.1   Cements covered by BS 12: 1996, 164
9.2.2   Sulfate-resisting Portland cement (SRPC), 168
9.2.2.1   Sulfate attack, 168
9.2.2.2   Sulfate-resisting Portland cement, 172
9.3   Cements with additional mineral constituents, 172
9.4   Pozzolanic materials, 173
9.4.1   Nature of the pozzolanic reaction, 174
9.4.2   Fly ash (pulverised fuel ash), 176
9.4.2.1   Composition, 176
9.4.2.2   Hydration, 179
9.4.3   Natural and related pozzolanas, 182
9.4.4   Silica fume (microsilica), 184
9.5   Blastfurnace slag and blastfurnace slag cements, 185
9.5.1   Composition, 185
9.5.2   Hydration, 188
9.6   Problems of specification of blended cements, 190
9.6.1   Sulfate-resistance of blended cements, 191
9.6.2   Influence of mineral additions in blended cements on the alkali-silica reaction, 194

10.   Admixtures and special cements   197
10.1   Admixtures, 197
10.1.1   Accelerators, 198
10.1.2   Retarders, 198
10.1.3   Water-reducing (plasticising) admixtures, 199
10.1.4   Air entrainment, 200

10.2   Oilwell cements, 200
10.3   Calcium aluminate cement (CAC), 201
10.4   Alkali-activated slag and aluminosilicate cements, 203
10.5   Calcium sulfoaluminate cements, 204
     10.5.1   Expansive and shrinkage compensated
            cements, 205
     10.5.2   Sulfoaluminate-belite cements, 206
     10.5.3   Practical considerations, 208

Appendix   Pore size distribution from an adsorption
            isotherm   209

Suggested further reading   211

Index   219

1. Introduction and composition of Portland cement

1.1 Introduction
1.1.1 Cementitious systems

A *cement* is a material which binds together solid bodies (*aggregate*) by hardening from a plastic state. This definition includes organic polymer-based cements. Apart from their use as adhesives, some larger scale use of polymers has developed, mainly in the USA, as binders for aggregate in a rapidly hardening patching material for damaged roads and bridge decks, for example. A monomer such as methyl methacrylate is polymerised and hardened *in situ*. However, the use of materials as expensive as these is very limited compared to the use of inorganic cements among which Portland cement is pre-eminent.

An inorganic cement functions by forming a plastic *paste* when mixed with water which develops rigidity (*sets*) and then steadily increases in compressive strength (*hardens*) by chemical reaction with the water (*hydration*). A cement which increases in strength even when stored under water after setting is said to be *hydraulic*.

The predominance of Portland cement, the major components of which are tri- and di-calcium silicates, in building and civil engineering is such that it is usually referred to simply as cement. Portland cements for special applications and binders in which Portland cement is blended with minerals such as pulverised fuel ash are considered in Chapter 9, while some unrelated special cements are described in Chapter 10. Four other cementitious systems with different modes of hardening merit brief mention here.

(a) Calcium sulfate (gypsum) plasters
A range of products exists, obtained by partial or complete dehydration of gypsum, e.g. plaster of Paris, Keene's cement. They harden by rehydration to form an interlocking mass of gypsum crystals:

$$CaSO_4 \cdot \frac{1}{2}H_2O \xrightarrow{\frac{3}{2}H_2O} CaSO_4 \cdot 2H_2O \qquad (1.1)$$

(b) Sorel cement
This hardens by the interaction of lightly fired magnesia with a concentrated solution of magnesium chloride to form a network of crystals of a basic chloride and magnesium hydroxide.

(c) Phosphate bonding
An aqueous phosphoric acid solution can be used to bond an oxide aggregate (e.g. alumina). An alternative 'one-pack' system containing a solid acid phosphate, for example $NH_4H_2PO_4$, merely requires the addition of water. The bond involves a glassy phosphate phase.

(d) Sodium silicate (water glass)
This produces a bond by decomposition of the silicate with polymerisation of the silica formed and carbon dioxide may be used to accelerate hardening. Spray dried sodium silicate can be used in a one-pack formulation.

Hardened gypsum plaster and Sorel cement are both degraded by exposure to an excess of water and are therefore used internally in buildings. Sorel cement, which produces a hard, abrasion-resistant flooring material, is so sensitive to moisture that it requires a protective coating of wax. Phosphate and sodium silicate systems are much more expensive than Portland cement but they are useful where rapid hardening or acid resistance merit the cost, or in special applications such as phosphate bonding in refractories and the use of sodium silicate in hardening foundry moulds.

1.1.2 Portland cement

A paste of Portland cement develops strength primarily by the hydration of the di- and tri-calcium silicates it contains. However, the chemical reactions of these compounds with water are far more complex than that for gypsum plaster. Because there are two products in the reaction (calcium hydroxide, sometimes referred to as portlandite, and an ill-defined calcium silicate hydrate) the process could be described as hydrolysis rather than hydration.

Concrete is a composite material produced by using cement to bind fine and coarse aggregate (sand and gravel or a crushed rock such as limestone or granite) into a dense coherent mass. The name Portland cement originated either because concrete resembled, or could replace, Portland stone. When smaller volumes of a more plastic material than concrete are required (for example between bricks) then a fine sand and cement mix (*mortar*) is used. An additional plasticiser such as hydrated lime or an organic wetting agent is usually added.

A cement–water paste (*neat paste*) is not normally used alone for economic and practical reasons. It would be too expensive, evolve too much heat by chemical reaction with the water if used in bulk, and when hardened undergo undesirably large volume changes when exposed to alternating dry and wet conditions in service (Chapter 8). However, for special applications such as the filling and sealing of small spaces in civil engineering constructions, neat paste (*grout*) is used.

The successful large-scale use of concrete depends principally on two aspects of the Portland cement–water reaction:

(*a*) there is a so-called *dormant or induction period* during which rates of hydration are low and for a substantial part of it a mix retains sufficient plasticity for proper placing and compaction;

(*b*) the hydration products formed after setting progressively fill the space previously occupied by water giving a dense structure with no undesirable overall volume changes.

The realisation of optimal engineering properties in concrete depends on the efficient use of these characteristics of cement. Poor compaction when placing or the use of too great an excess of water over that necessary for the hydration reactions produces porous concrete. This may be strong enough but have low resistance to abrasion, freeze–thaw disintegration, and attack by aggressive environments. Furthermore, it will be too permeable to ensure adequate protection of the steel used in *reinforced concrete* to provide tensile and flexural strength. The poorly protected steel corrodes and expansive iron oxide formation eventually leads to cracking and disintegration of the concrete.

1.1.3 Historical aspects

Both concretes and mortars based on lime were used in antiquity. The Romans recognised the importance of compaction to produce dense, durable material. Examples of their success are the Pantheon in Rome and the Pont du Gard aqueduct in southern France, both of which have survived intact. Roman mortar contained hydrated (slaked) lime produced from limestone by calcination and hydration of the quicklime.

$$CaCO_3 \longrightarrow CO_2 + CaO \xrightarrow{H_2O} Ca(OH)_2 \tag{1.2}$$

The hydrated lime was intimately mixed with sand and a fine material of volcanic origin. This mortar hardened as a compacted mass in the presence of water by interaction of the volcanic material with lime to form hydrated calcium silicates similar to those found in hardened

Portland cement paste. Natural and artificial materials which undergo such reactions with lime are said to be *pozzolanic* after Pozzuoli in Italy where a natural example is found.

The skills of the Romans were lost in the Dark Ages and inferior mortars which hardened mainly by carbonation of lime were used. Eventually, the effective use of pozzolanas was rediscovered and sources in Holland, for example, were used in Northern Europe.

The modern development of hydraulic cements can be said to stem from the discovery by Smeaton (1793) that limestone containing clay on calcination gave a lime which hardened under water. This *hydraulic lime* did not disintegrate on hydration as do the *fat* (pure) *limes* required for the metallurgical and glass industries. Smeaton used a hydraulic lime and a pozzolana in building the second Eddystone lighthouse.

Towards the end of the eighteenth century it was observed that when some argillaceous limestones were calcined they gave a hydraulic cement and the naturally occurring mix of the appropriate materials became known as *cement stone* or *rock*. By modern standards these products were only lightly calcined so as not to be too greatly densified by sintering and therefore easily ground.

Joseph Aspdin, who obtained a patent in 1824, is usually regarded as the inventor of modern Portland cement, although his product, made by *burning* chalk with clay in a lime kiln and grinding the sintered product (*clinker*), was much inferior to the present-day material. Two fundamental developments have improved on the early product: the introduction of gypsum, added when grinding the clinker, to act as a set retarder (*ca.* 1890) and the use of higher burning temperatures to allow the production of higher lime content silicates which are necessary for more rapid strength development in concrete. Improvements were aided by the gradual replacement of vertical shaft kilns by the rotary kiln and the introduction of the ball mill for the grinding of cement (both from *ca.* 1890).

The second half of the twentieth century has seen the reduction in consumption of fuel and electric power become a major objective. Fuel saving measures have included the almost universal replacement of an aqueous slurry of the raw materials as kiln feed by drying, grinding and blending them to produce a powder feed for a heat exchanger/kiln system. In addition, economies of scale have resulted in the construction of larger plants, around 1 million tonnes per annum now being common. Most recently, significant reductions in electric power consumption have been achieved where roller mills have either replaced or supplemented ball milling of raw materials. Savings have often been offset, however, by the need to use harder raw materials and to grind the cement finer to meet market requirements. Increasing availability and sophistication of instrumental control of kiln feed composition and burning have been exploited in the production of cement with a quality as consistent as is possible from virtually as-dug raw materials. This overall quality objective is aided by a determination in the laboratory of the optimum clinker composition and microstructure that can be produced from a given set of raw materials.

1.2   Composition of Portland cement
1.2.1   Nomenclature

Portland cement clinker is produced by burning a mix of calcium carbonate (limestone or chalk) and an aluminosilicate (clay or shale) and then grinding the product with approximately 5% gypsum to produce cement. Typical compositions are given in Table 1.1 with compound contents calculated by the Bogue procedure (Section 1.2.3). The significant levels of iron oxide in a grey cement are derived from the clay, as are the much lower levels of alkalis. The relatively small amount of white cement manufactured requires raw materials with a very low iron content. The great majority of the grey cement marketed in the UK was formerly designated *Ordinary Portland Cement* in British Standard BS 12. It is now included in the designation CEM I in the current BS 12: 1996

Table 1.1.   Composition of Portland cement

	Cement*	Clinker*		Cement	Clinker	
	Grey: %	Black: %	White: %	Grey	Black	White
SiO_2	19 – 23	21.7	23.8	LSF%[‡] 90–98	98.4	97.2
Al_2O_3	3 – 7	5.3	5.0	LCF%[‡] –	96.2	93.8
Fe_2O_3	1.5 – 4.5	2.6	0.2	S/R 2–4	2.7	4.6
CaO	63 – 67	67.7	70.8	A/F 1–4	2.0	25
MgO	0.5 – 2.5	1.3	0.08	C_3S% –	65.4	59.4
K_2O	0.1 – 1.2	0.5	0.03	C_2S% –	12.9	23.5
Na_2O	0.07 – 0.4	0.2	0.03	C_3A% –	9.6	12.9
SO_3	2.5 – 3.5[†]	0.7	0.06	C_4AF% –	7.9	0.6
LOI	1 – 3.0[†]	–	–			
IR	0.3 – 1.5[†]	–	–			
Free lime[§]	0.5 – 1.5	1.5[§]	2.5[§]			

```
*      cement — usual range; clinkers — examples used in text
†      upper limits in BS 12: 1996
‡      lime saturation and combination factors (Section 2.2)
§      also included in total CaO
LOI  loss on ignition (CO₂ + H₂O) typically 0.8–1.8%
```
* cement — usual range; clinkers — examples used in text
[†] upper limits in BS 12: 1996
[‡] lime saturation and combination factors (Section 2.2)
[§] also included in total CaO
LOI loss on ignition (CO_2 + H_2O) typically 0.8–1.8%
IR insoluble residue — usually siliceous and typically <1%
S/R silica ratio % SiO_2/(%Al_2O_3 + %Fe_2O_3)
A/F alumina ratio %Al_2O_3/%Fe_2O_3

and the European, currently voluntary pre-standard, ENV 197. Following the latter, Type I is subdivided into *strength classes* (Table 6.1) on the basis of a minimum mortar crushing strength required after 28 days' prescribed curing. That for the common product is 42.5 N/mm^2. In the literature, the word 'ordinary' is still used in the normal adjectival sense for this cement and the abbreviation opc is frequently seen. The equivalent in the American Society for Testing and Materials classification is ASTM Type I.

It is customary to report chemical analyses as contents of oxides since the compounds have empirical formulae given by the addition of the oxide formulae. For example, tricalcium silicate Ca_3SiO_5 is equivalent to $3CaO + SiO_2$. While such relationships are useful for calculation of quantities, they naturally tell us nothing about the structural nature of the calcium silicate. A form of shorthand known as the 'cement chemist's notation' is used to simplify formulae. As in refractory technology, single letters replace the usual oxide formulae.

Oxide	CaO	SiO$_2$	Al$_2$O$_3$	Fe$_2$O$_3$	H$_2$O	Na$_2$O	K$_2$O	SO$_3$	MgO	CO$_2$
Symbol	C	S	A	F	H	N	K	$\bar{S}$	M	$\bar{C}$

A compound such as tricalcium silicate is then written as C_3S. More complex compounds are sometimes written in a hybrid formulation to emphasise some aspect of their composition. For example, the sulfoaluminate, ettringite (Section 7.2.3) may be seen represented by $C_3A.3CaSO_4.32H_2O$ rather than $C_6A\bar{S}_3H_{32}$.

1.2.2   High-temperature phase equilibria

In this elementary account of the phase equilibria involved in cement making we will ignore all but the principal oxides. The relevant features of the C–S–A–F quaternary system were determined by Lea and Parker (1934) but the principles used by the cement chemist to produce a suitable mix of the oxides from the available raw materials can be illustrated by using the ternary C–S–A system (Rankin and Wright, 1915; Osborn and Muan, 1960). The relevant portion of the diagram is shown in Fig. 1.1 with temperatures for $C_{12}A_7$ melting modified according to Glasser and Williamson (1962). It applies to a white Portland cement clinker in which the iron content is negligible and the composition corresponding to that given in Table 1.1 is marked W.

The phase changes involved in equilibrium cooling from a total melt of compositions lying in the primary phase fields of C_2S, C_3S and lime,

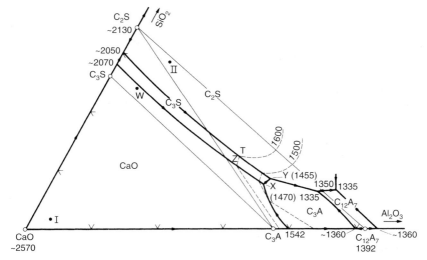

*Fig. 1.1.   High lime region of the CaO–SiO₂–Al₂O₃ phase diagram. Composi-
tions: W—white clinker (Section 1.2.2); I, II—intermediate compositions in
lime- and silica-rich regions of a raw materials mix (Section 2.3.1)*

for a section through W parallel to the Ca–SiO₂ edge, can be deduced
from the general treatment of a ternary system with an incongruently
melting compound. W lies in the composition triangle of C_3S, C_2S and
C_3A (tricalcium aluminate) so that if equilibrium is established at
burning temperature and maintained during cooling then only these three
compounds should be obtained. However, C_3S melts incongruently
(Welch and Gutt, 1959) and W lies within the primary phase field of
lime. During cooling this lime forms C_3S by reaction with the relatively
lime-deficient liquid, the composition of which follows the C_3S–lime
boundary toward the invariant point X with composition: 59.7%C,
32.8%A, 7.5%S.

For the compositions on the low-lime side of the line joining the C_3S
composition to the invariant point X, the combination of the lime is
completed during equilibrium cooling before this temperature (1470°) is
reached. For any composition on the high-lime side of this line, solid lime
remains when the invariant temperature is reached. At this point C_3A
begins to crystallise with a tendency to encapsulate the lime, so that
equilibration as cooling continues is slow. Since the amount of
uncombined lime (*free lime*) in a clinker must be low (to remove the
risk of unsoundness in concrete — Section 6.6), and because of the
difficulty in a commercial process of maintaining equilibrium during
cooling, it is necessary to use a mix of oxides which theoretically can
equilibrate to zero free lime at temperatures above the invariant point.

Consequently, compositions richer in lime than those lying on the line C_3S–X are avoided and those on the line are said to have a *lime saturation factor* of 1.00 or 100%. In practice, cement clinker production only involves partial melting so that equilibrium is approached in a relatively slow and complex manner which will be considered in Section 2.3.1. At 1600° composition W will consist of about 19% liquid of composition T in equilibrium with a mixture of C_3S and C_2S. Thus, if the aim is to produce a clinker with an equilibrium solid uncombined lime content of zero *at the clinkering temperature*, then W is on the correct side of the line joining the C_3S composition to the 1600° isotherm at Z. The position of this line, which virtually represents the upper practical temperature limit for clinkering, is so similar to that of C_3S–X that a single line defines 100% lime saturation whichever temperature is chosen. Finally, it should be noted that if the liquid at 1470° were cooled without any interaction with solids then formation of $C_{12}A_7$ would be expected. However, it is very rarely observed and with the slowest rates of cooling attained in commercial production the liquid may actually take lime from previously formed C_3S.

In the quaternary system for grey Portland cement, compositions lie in a tetrahedron with an aluminoferrite as the fourth compound, having a composition approximating to C_4AF (Section 1.3.4). In the quaternary diagram a plane corresponding to 100% lime saturation contains both the C_3S–X line in the C–S–A base and the composition C_4AF. Lea and Parker gave a formula for *proportioning* the oxides to this lime saturation, the application of which to typical raw materials will be considered in Section 2.2. For a more detailed discussion of these phase diagrams the reader is referred to Taylor (1997) and Macphee and Lachowski (1998).

1.2.3   Compound composition of Portland cement

The potential compound composition of the clinkers in Table 1.1 could be determined from the phase diagrams but usually a simple computational method given by Bogue (1929) is employed. The term potential was used because it was assumed that the set of oxides is brought to equilibrium and kept in equilibrium during cooling, a situation which is not fully realised in practice.

In the calculation all of the iron oxide is assumed to be present in an aluminoferrite solid solution of composition C_4AF and all of the alumina not required to satisfy the iron in this ferrite is considered to be present as C_3A. The lime remaining, after that required for these compounds is deducted from the combined lime, is then proportioned between C_3S and C_2S. The combined lime is simply the total found by chemical analysis minus the free lime which can be determined by its extraction with ethanediol (Section 4.2).

The general solution to the calculation is expressed in the following equations (Bogue, 1955) in which the chemical formulae represent the mass % of each oxide:

$$C_4AF = 3.04Fe_2O_3$$
$$C_3A = 2.65Al_2O_3 - 1.69Fe_2O_3$$
$$C_2S = 8.60SiO_2 + 1.08Fe_2O_3 + 5.07Al_2O_3 - 3.07CaO$$
$$C_3S = 4.07CaO - 7.60SiO_2 - 1.43Fe_2O_3 - 6.72Al_2O_3$$

If the Bogue calculation is used with the analysis of a cement rather than a clinker then the gypsum (ca. 5%) is inevitably included in the analysis. It therefore contributes to the total lime and the extent of this is computed from the total sulfate determined by analysis. For example, the lime equivalent to a typical SO_3 content of 2.6% is 1.8% (2.6 × 0.7). Since some of this sulfate is derived from the clinker component of the cement and may not be combined with lime, this procedure introduces a small error.

The Bogue compositions obtained by applying the calculation to the white and grey cement clinkers are given in Table 1.1. The totals differ from 100% because minor components have been ignored. The accuracy of the results is further limited not only by the failure of a system to remain in equilibrium but also by solid solution effects, which result in the four main clinker phases having compositions significantly different from those in the simple C–S–A–F system. The phase composition of a clinker may be determined by quantitative application of light microscopy or X-ray diffraction and the composition of individual phases by X-ray microanalysis (Chapter 4). Taylor (1997) gave typical compositions for clinker phases and described a modification of the Bogue procedure to allow for solid solution.

1.3   Polymorphism and solid solution in clinker compounds

All the major phases in commercial Portland cement clinker exist as multi-component solid solutions. Examples are numerous and both major and minor constituent elements are involved. In addition, polymorphism (that is the existence of different crystalline forms or phases with the same nominal chemical composition) is exhibited by the silicates, for example, and the stability of polymorphs is influenced by relatively low levels of solid solution. Before giving a brief account of the structures of the clinker compounds, it is useful to recall the main features of solid solution relevant to cement.

The term solid solution does not encompass inclusions which are observed in clinker by optical microscopy. It involves a distribution of the solute ions either by substitution of them for ions of the host crystal or by their location in normally unoccupied interstitial sites. These processes may produce vacancies at normally occupied sites to maintain electroneutrality, when substitution of a cation M^{2+} by M^{3+} occurs, for example. A charge balance in substitution may also be maintained by the creation of free electrons or positive holes in the conduction or valence bands of the solid, respectively. These may significantly increase chemical reactivity and also produce a deep colour.

The defect concentration in a solid is made up of substitutional, point defects and line defects such as edge and screw dislocations. All except substitutional defects are present in pure solids. The number of defects per unit volume of a solid depends on its chemical composition and the thermal and mechanical treatment it has received. In addition, *extended defects* involving rows or blocks of atoms occur and those associated with changes in crystal symmetry accompanying cooling from high temperatures are relevant here. Such defects may involve a significant fraction of the total number of atoms in a small crystal and the accompanying strain produces an increase in the Gibbs free energy of the solid. This can result in enhanced reactivity in hydration with the possibility of accelerated localised reaction of atoms at sites of high strain. Timashev (1980) reported Russian work in which microscopic determination of etch pit density showed 1/3 to 1/2 of the surface of C_3S and 1/6 to 1/3 of that of C_2S to be occupied by dislocations.

Solid solution often produces a change in the unit cell size of a crystal lattice which depends on the concentration of solute ions and which can be determined by X-ray diffraction. Above a critical concentration there may be a change in unit cell type (symmetry). Solid solution may stabilise a high-temperature polymorph so that on cooling it does not invert to the form stable at room temperature. This effect frequently depends on the rate of cooling: rapid cooling favours the retention of larger amounts in solid solution, while slow cooling favours ex-solution of foreign ions and phase inversion. At room temperature the thermal energy available may not be sufficient for the nucleation of the stable phase, although this may sometimes be induced by mechanical shock. A metastable form may also persist if the crystallite size is small because the probability of nucleation of the stable phase is then low.

Solid-solution compositions in cement clinker are usually expressed in chemical analyses as if oxides rather than elemental ions had gone into solution. This is acceptable because the crystal structures of the cement minerals are based on a framework of O^{2-} ions into which cations are packed. For electroneutrality each cation is accompanied by the number of O^{2-} ions in its oxide formula. Elements such as silicon and sulphur are treated as hypothetical cations (Si^{4+}: S^{6+}) in describing structures.

In characterising the interaction of a particular phase and related solid solutions with water it is necessary to distinguish between the rate of the chemical reaction in hydration and the development of strength in a mortar or paste hardening under water. The term *hydraulicity* is usually reserved for the latter and the two evaluations may differ significantly. For example, C_3A is the most reactive constituent of clinker but pastes of it develop little strength.

1.3.1   Tricalcium silicate (alite)

This is an orthosilicate containing discrete SiO_4^{4-} tetrahedra and O^{2-} ions as anions, so that its formula may be written $Ca_3(SiO_4)O$. Three triclinic (T), three monoclinic (M) and rhombohedral (R) forms exist, symmetry increasing with increasing temperature as expected. Phase-inversion temperatures have been established both by X-ray diffraction and differential thermal analysis. Structure changes are small and so therefore are the thermal effects.

$$620° \qquad 920° \qquad 980° \qquad 990° \qquad 1060° \qquad 1070°$$
$$T_1 \longleftrightarrow T_2 \longleftrightarrow T_3 \longleftrightarrow M_1 \longleftrightarrow M_2 \longleftrightarrow M_3 \longleftrightarrow R$$

C_3S is stable (with respect to $C_2S + C$) between about 1250° and 1800° and melts incongruently at 2150°. The high temperature forms can be stabilised at room temperature by solid solution of the impurities present in the raw materials used in cement manufacture and the impure C_3S present in commercial clinkers is referred to as *alite*. Alite is usually monoclinic (M1 or M3), occasionally rhombohedral, or triclinic. The most frequently found foreign ions are Mg^{2+}, Al^{3+} and Fe^{3+} with smaller amounts of K^+, Na^+ and SO_3. Maki (1994) related the crystalline form of alite in clinkers to the level of MgO and SO_3 present. Transition metals such as chromium and manganese (oxidation state variable) are also usually present at low levels and have significant effects on colour in a white cement. The different forms of alite do not differ greatly in hydraulicity. The formation of solid solutions increases the percentage of alite significantly above that calculated for C_3S by the Bogue method. For example, Mg^{2+} can replace Ca^{2+} in alite which increases the apparent C_3S level but decreases that of C_2S.

1.3.2   Dicalcium silicate (belite)

Dicalcium silicate is also an orthosilicate and, as with the alites, differences between polymorphs principally involve the orientation of the SiO_4 tetrahedra. Some transformations occur over a temperature range which is different for rising and falling temperature. There are four principal polymorphs which have been designated α, α', β and γ. Of these α' has two very similar forms distinguished as α'_H and α'_L. There are discrepancies in the published literature on the transformations which occur, partly as a result of the more recent findings that time at higher temperature can affect the lower temperature transformations and that transformations are crystallite size dependent.

A simplified scheme is:

$$
\gamma \xleftarrow{} \beta \underset{630°}{\overset{680°}{\rightleftarrows}} \alpha'_L \xleftrightarrow{1160°} \alpha'_H \xleftrightarrow{1425°} \alpha \xrightarrow{2130°} \text{congruent melting}
$$
$$
\underset{800-850°}{\underline{\quad < 500° \quad}}
$$

Differences in hydraulicity are considerable. The γ form has an orthorhombic, olivine, structure (specific gravity 2.94) and is essentially non-hydraulic although a paste does hydrate slowly (Bensted, 1978). The common form in commercial clinker is the monoclinic βC_2S (specific gravity 3.30). This has a strongly distorted low-temperature K_2SO_4 structure, while α'_H-C_2S has a high-temperature K_2SO_4 structure. (Note the formal similarity to Ca_2SiO_4).αC_2S has a glaserite ($NaK_3(SO_4)_2$) structure.

The impure form of C_2S in commercial clinker is referred to as *belite*, usually β but occasionally α' is found. Belite is stabilised with respect to γC_2S by foreign ions in solid solution. If the β–γ phase inversion occurs in commercial cement clinker during cooling then the large specific volume change leads to disintegration (*dusting*) of the clinker with serious consequences for the efficiency of the cooler and, since γC_2S is non-hydraulic, for the quality of the product. Fortunately this occurrence is very rare because alkalis, in particular K^+, are usually available in sufficient quantity to stabilise βC_2S.

$\alpha' C_2S$ is said to be more hydraulic than βC_2S but solid solution of phosphate ions reverses this. αC_2S stabilised at ambient temperature by large amounts of $Ca_3(PO_4)_2$ was found to be non-hydraulic by Welch and Gutt (1960). In general the proportions of foreign ions taken into solid solution are higher than in alite. Cations include Al^{3+}, Fe^{3+}, Mg^{2+}, K^+ and anions include SO_4^{2-} and PO_4^{3-}. Those ions which excessively stabilise C_2S at high temperatures during clinkering, e.g. phosphate, inhibit the formation of C_3S and amounts greater than about 0.3% must be avoided (Nurse, 1952). No link between impurity content, dislocation density and reactivity was found for the different forms of C_2S by Boikova (1978).

1.3.3   Tricalcium aluminate

Tricalcium aluminate is cubic and does not exhibit polymorphism when pure. Mondal and Jeffery (1975) showed that it contained six-membered kinked rings of AlO_4 tetrahedra ($Al_6O_{18}^{18-}$). It reacts vigorously with water but pastes develop little strength. In cement clinker Fe^{3+}, Mg^{2+}, Na^+, K^+ and Si^{4+} are found in solid solution, but only the alkalis change the symmetry of the structure. The solid solution containing sodium is that most commonly encountered and its reaction with water is less vigorous than that of pure C_3A. Takeuchi *et al.* (1980) made a detailed study of the effects of sodium on the symmetry and inter-atomic spacings in C_3A solid solutions. Their results can be summarised as follows.

%Na_2O
< 1.0   Cubic
1.0–2.4   Cubic with different space group
2.4–3.7   Orthorhombic + cubic forms
3.7–4.6   Orthorhombic
4.6–5.7   Monoclinic

Solid solution involves the substitution of Ca^{2+} by Na^+ with a second Na^+ going into a vacant site at the centre of a six-membered ring of AlO_4 tetrahedra. The effect of the new charge distribution is an expansion of the Ca–O polyhedra while the ring contracts. The net effect is a very small decrease in cubic cell size: from 1.5263 nm in C_3A itself to 1.5248 nm at 2.4% Na_2O. Site occupation by sodium is affected by the presence of silica in solid solution which extends the range for sodium above 5.7 towards the composition NC_8A_3, which is not, however, a specific compound.

1.3.4   Calcium aluminoferrite: the ferrite phase

In the C–A–F system, C_2F is stable under ambient conditions but C_2A is stable only at high pressures (Glasser and Biggar, 1972). In the solid solution series $C_2\,A_xF_{1-x}$ the limit to x at normal pressures is ca. 0.7. C_2F is orthorhombic, layers of FeO_6 octahedra alternating with layers of FeO_4 tetrahedra with Ca^{2+} ions packed between them. With values of x up to 0.33, Al^{3+} replaces Fe^{3+} in tetrahedral sites, while at higher levels of replacement the additional Al^{3+} goes increasingly into octahedral sites.

For the aluminoferrite phase in Portland cement the formula C_4AF ($x = 0.5$) is often assumed, in the Bogue calculation for example. In fact the A/F ratio of the ferrite phase in commercial clinker varies and it contains significant proportions of other elements in solid solution. The most usual are Mg^{2+}, Si^{4+} and Ti^{4+} and Taylor (1997) gave a typical composition as: $Ca_2AlFe_{0.6}Mg_{0.2}Si_{0.15}Ti_{0.05}O_5$. Variability in composition is found not only between different clinkers but also within a single sample.

This phase is the only strongly coloured compound in the quaternary system. In thin section it is yellow-brown in transmitted light. The black colour and associated semiconductivity of the ferrite in Portland cement are said to result from the combined effects of solid solution (in particular Si^{4+}) and cooling in an oxidising atmosphere after firing, suggesting the formation of some Fe^{4+} ions (Scrivener and Taylor, 1995; Maki *et al.*, 1995). Reactivity with water is moderate and increases with increasing A/F ratio, but hydraulicity is slight. However, if this compound is formed by sintering at lower temperatures than those needed for partial melting (clinkering) then it is more reactive and its hydraulicity is enhanced.

1.4   Minor clinker compounds

In addition to the four principal compounds, commercial Portland cements may also contain several minor constituents as distinct phases. Magnesia may be retained in solid solution or, if present at a level of above about 1.5%, be present as periclase. A substantial proportion of the sodium present goes into solid solution in the main phases, while the remainder and over 90% of the potassium combine with the sulfur in the raw materials and fuel. Some sulfate goes into the silicates but it is largely present as alkali or alkali calcium sulfates. Depending on the relative proportions of sulfate and alkali in the clinker, various combinations of arcanite (K_2SO_4), aphthitalite ($K_3Na(SO_4)_2$), calcium langbeinite ($Ca_2K_2(SO_4)_3$) and, rarely, anhydrite ($CaSO_4$) can be formed.

2. Raw materials

2.1 Raw materials

It was seen in Chapter 1 that the principal oxide of the quaternary system in which the Portland cement compositions lie is CaO. In practice this is derived from naturally occurring calcium carbonate (calcite), which makes up approximately 75–80% of the *raw* materials *mix* (*meal*) used in making cement clinker. While this *primary* calcareous raw material may contain significant amounts of the other oxides of the quaternary system as impurities, a *secondary* raw material is normally required to provide the bulk of these. This is usually a clay (argil) or shale.

The location of a cement works is usually determined by the availability of adequate supplies of the raw materials within a reasonable distance of each other. The quality, uniformity, quantity and likely problems of extraction (*winning*), that is the amount of overburden to be removed and form of bedding, are established by geological survey involving drilling and the analysis of the cores obtained. The world distribution of suitable raw materials is far from uniform, the UK being atypical in containing an abundance of the sedimentary strata necessary.

2.1.1 Calcareous component

The minerals used in cement manufacture cover the complete range of geological eras including all types of limestone and the metamorphosed form, marble. The differences of most practical interest are crystal size, porosity and hardness. Those limestones which are porous and easily crushed usually contain a significant amount of water; levels approach 25% in the lower Thames Valley Cretaceous chalks which contain unrecrystallised skeletal relicts (Fig. 2.1). Crystal growth and densification has occurred in most of the older limestones to an extent dependent on the temperature, pressure and humidity they have experienced.

The proportion of argillaceous and siliceous material in a limestone may be sufficient in quantity and with a suitable distribution to make up a natural cement stone requiring little more than burning. This is true of the

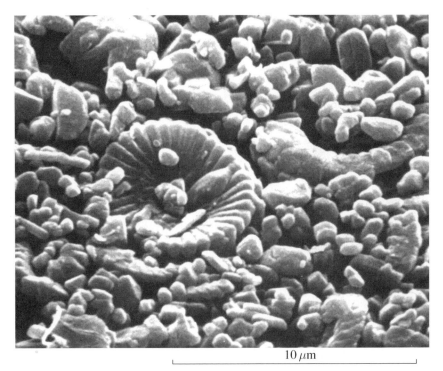

10 μm

Fig. 2.1.   Coccolith in a very porous chalk (courtesy of Blue Circle Industries)

rock in the Lehigh Valley (USA) and of the Cambridgeshire marls (calcareous muds), although such materials are no longer of commercial significance. However, a siliceous limestone may be used as a secondary raw material so that a kiln feed is made up of a blend of a high and a low carbonate stone.

Common minor impurities, the quantity and distribution of which must also be established by the survey of the deposit, are magnesia (as magnesite); and minerals containing lead, zinc, copper, barium, fluoride and phosphate. Magnesia levels must not be too high for the reasons discussed in Section 6.6. Metallic minerals occur in veins or rakes and the stone containing these can be rejected during quarrying if the amount is not too great. Even at levels of only a few tenths of one percent, heavy metals inhibit the setting of cement. Fluorspar (CaF_2) is a useful mineraliser aiding combination in the cement kiln, but if fluorine is present at above about 0.2% in cement then setting is retarded. If more than about 0.3% of phosphate is present then its ability to stabilise C_2S and inhibit C_3S formation becomes a problem.

Flint, which commonly occurs in chalk, while not chemically undesirable, is somewhat unreactive and expensive to grind. However,

Fig. 2.2. Quarrying chalk using a bucket excavator and mobile crusher (courtesy of Blue Circle Industries)

in the production of a slurry of the chalk in water, flints can be used as media to break the chalk down (*autogenous grinding*) and can then be removed mechanically. For white cement a source of carbonate low in iron, manganese and chromium is required since these elements have a strong influence on its brightness.

The method of winning limestone depends on its hardness. Soft chalks and marls are scraped from the quarry face (Fig. 2.2) and crushed in the quarry before transport to the blending plant. Hard material requires blasting and one or more stages of crushing before it is ground. Variations in composition are accommodated by blending in a stockpile and by planned development of the quarry faces, using the geological survey and any additional compositional trends indicated by regular chemical analysis as material is extracted. Beneficiation other than simple rejection of unsatisfactory material is rare, although occasionally flotation is used.

2.1.2 Argillaceous component
In cement making, overall chemical composition of a clay or shale is the chief consideration. The clays used generally contain a mixture of two-layer minerals of the kaolinite group and three-layer minerals of the illite and montmorillonoid (swelling clays) groups. Free silica (usually quartz) may also be a major constituent. Alkalis are derived partly from the exchangeable cations in clays, but mainly from any micas and feldspars

present. Calcite, iron pyrites, gypsum and organic matter are other common minor components. Some shales have to be rejected because the sulphur and/or alkali levels are too high. The chloride content must be low and this limits the use of estuarine muds. A number of other elements are derived from clay minerals because of a degree of substitutional solid solution in place of Si^{4+} and Al^{3+}. The major substituent is usually iron but magnesium, chromium and manganese are also found.

The quantitative determination of clay minerals by X-ray diffraction is particularly difficult since the interstratification of different types to form mixed-layer clays is common. The more crystalline components such as kaolinite, quartz and calcite are readily detected, but even the presence of poorly crystalline three-layer clays is difficult to confirm in natural mixtures. Swelling clays are of practical interest because they can seriously increase the viscosity of the aqueous slurry used in the wet process of cement manufacture (Section 3.3.2). In contrast, they have beneficial effects in the semi-dry process (Section 3.3.4). Their presence is most easily revealed by X-ray diffraction, using the fact that the 9.2 Å spacing in a dried sample is expanded to 13.6 Å after it has been treated with ethanediol. To assess the potential behaviour of a clay in cement manufacture more quantitatively, determination of its cation-exchange capacity is useful. Methylene blue is a convenient substance for this determination but care is needed in interpreting the results as they may be influenced by organic matter in the clay.

The ideal formulae for the prototype clay minerals define the limits for the silica ratio $(S/A + F)$ which can be derived from the clay in a raw materials mix.

	Formula weight	S/(A + F)
Kaolinite $Al_2(OH)_4Si_2O_5$	258	1.18
Pyrophyllite $Al_2(OH)_2(Si_2O_5)_2$	360	2.36

If a secondary raw material contains a high proportion of kaolinite then the silica ratio of a mix with limestone may be too low to yield an adequate proportion of calcium silicates in the corresponding clinker. Frequently, there is enough free silica in the clay and/or limestone to compensate for the kaolinite but, if not, an additional component such as sand is used.

Methods of extracting depend on the degree of compaction experienced by the deposit. Soft, high moisture content clays are mechanically excavated by scraping and transported from the quarry by conveyor systems. Hard shales are blasted and crushed before being stockpiled.

2.1.3   Multicomponent mixes

Primary and secondary raw materials may be available with a range of compositions in a given locality and several may be needed to give a blend with a composition capable of being burned to an acceptable clinker. The use of multicomponent mixes increases both as prolonged extraction from a quarry reduces the availability of the highest quality material and as the market demands greater consistency in the product. It may in some cases be necessary to 'import' a component to get the optimum composition, although this is only possible on a limited scale unless a special cement is being made. Examples of such materials are sand, pulverised fuel ash (essentially burnt clay originally associated with coal deposits), iron pyrites cinders (iron oxide), siliceous loams, diatomaceous earths and china clay (for white cement). A low alkali content ash may provide a substitute for a high alkali or high sulfur shale and the residual carbon in an ash may be sufficient to contribute significantly to fuel economy.

2.1.4   Gypsum

Gypsum, while not a raw material in the kiln feed, is used in Portland cement to regulate set. It is added to clinker at the cement grinding stage (Section 5.1), usually at a level of 5–6%, depending on its purity. The addition is calculated from the sulfate content of the two materials, the gypsum generally being used as received. It may contain anhydrite, and clay, quartz and calcite as impurities, at levels which vary widely with source. Deposits may be mined or quarried and separation of the purest material is necessary for white cement. A material too rich in natural anhydrite (> about 70%) is unsuitable for use alone because this mineral dissolves too slowly in water to retard setting sufficiently.

In those parts of the world where adequate natural deposits of gypsum are unavailable, *chemical gypsums* may be employed. These are by-products of a number of chemical processes in which at some stage sulfuric acid is neutralised by the addition of hydrated lime. Depending on temperature and solution concentration, anhydrite or the dihydrate, gypsum, separates. This microcrystalline form of anhydrite is much more rapidly soluble than natural anhydrite and so can be used directly. However, the use of these by-products depends on their not being too contaminated by residues of a product of the chemical process (such as phosphate ions in phosphoric acid production) which will affect the setting and/or hardening of a concrete.

2.1.5   Coal ash

When coal is used to fire a kiln, ash is absorbed into the clinker produced and this must be allowed for when calculating the proportions of the raw materials to be used in the kiln feed. The ash produced by burning coal from a UK source is derived from the shales associated with coal measures and is, therefore, an argillaceous/siliceous component. To assist in controlling the composition of a clinker, it is desirable that coal from one source, with a steady ash content and composition, is used at a given cement works. Coal must be adequately ground so that ash is carried well back into a kiln system where it can be uniformly incorporated. Inadequate preparation of coal may result in coarse ash falling onto the surface of already formed nodules, with the clinkering processes failing to eliminate the major chemical imbalance between the core and the skin of a nodule.

Ash contents of bituminous coals extracted in the UK normally lie in the range 5–20%, but in some countries brown coals with over 30% ash are used. The chemical composition of coal ash varies considerably worldwide. Although using oil or natural gas removes the ash problem, choice is based on cost, except in the case of white cement for which coal ash has too high an iron content. Both coal and oil contribute some sulfur which appears as sulfates in the clinker (Section 1.4).

2.2   Proportioning of raw materials

As explained in Section 1.2.2, the raw materials are blended in a way determined by the quaternary C–S–A–F phase diagram. The composition and reactivity of the minerals used will determine how closely the ideal, which is a high silicates content with a low free lime, can be approached.

2.2.1   Calculation of the composition of a kiln feed

Although the mathematics involved is elementary it is often unfamiliar and the calculation is best illustrated by an example. Analyses of the chosen materials are given in Table 2.1. As the four oxides are present in the chalk and the clay we have to calculate (1) the excess of lime available from the chalk and (2) the additional lime needed to saturate the acidic oxides in the clay at a chosen level. The ratio of these two gives the proportions of chalk and clay required in the mix.

Lea and Parker (1935) derived a formula from the plane of 100% lime saturation in the quaternary which makes possible the calculation of the lime required for saturation of the other oxides at the clinkering temperature:

$$CaO = 2.80 SiO_2 + 1.18 Al_2O_3 + 0.65 Fe_2O_3 \qquad (2.1)$$

in which chemical formulae are used to represent mass % of the oxide.

The lime saturation factor (LSF) for any mix of raw materials is then given by the ratio of the lime available to the lime theoretically required:

Table 2.1. Chemical analyses of raw materials, derived mixes and clinkers

	Chalk: %	Clay: %	Loam: %	Ash: %
S	2.5	50	84	48
A	0.5	22	6	29
F	0.2	9	3	10
C	54.0	2.5	1	8
Res*	42.8	16.5	6	5

	Mixes			Clinkers	
	A %	B %	C %	C %	D %
S	13.09	16.05	14.53	22.73	23.24
A	5.30	1.42	3.41	5.36	5.83
F	2.17	0.67	1.44	2.25	2.40
C	42.52	45.01	43.74	68.45	67.24
Res*	36.92	36.85	36.88	1.22	1.30
LSF	95.9	96.0	96.0	96.0	91.4
S/A + F	1.75	7.68	3.0	3.0	2.8

* Residuals, including loss on ignition except for ash and clinkers

$$LSF = \frac{CaO}{2.80SiO_2 + 1.18Al_2O_3 + 0.65Fe_2O_3} \tag{2.2}$$

This ratio is usually expressed as a percentage and it may also be applied to the analysis of a clinker or a cement, although for the latter the lime present as $CaSO_4$ is deducted from the total lime in the analysis, as was necessary in the Bogue calculation (Section 1.2.3).

Suppose that the reactivity of our raw materials after grinding, the amount of which is limited by costs, leads us to predict that if an LSF of 96% is exceeded the uncombined lime will be too high after burning for the residence time anticipated in the kiln. The lime required to saturate the acidic oxides to this level is:

$$0.96(2.80SiO_2 + 1.18Al_2O_3 + 0.65Fe_2O_3) \tag{2.3}$$

Equation (2.3) is applied to the analyses of both the chalk and clay in Table 2.1 with the following results:

lime required for acidic oxides in chalk $= 7.41$
∴ lime available to combine with clay $= 54.0 - 7.41 = 46.59$
lime required for acidic oxides in clay $= 164.9$
∴ net lime required for clay $= 164.9 - 2.5 = 162.4$.

Thus the ratio of chalk to clay to balance the two is $162.4/46.59 = 3.49$. This ratio gives the composition of Mix A in Table 2.1. If the silica ratio of this mix is considered to be too low to produce the proportion of silicates needed for good strength development, it can be raised by introducing a siliceous loam. It is first necessary to calculate the proportions of chalk and loam in a Mix B (Table 2.1) which would have an LSF of 96%, as only then can we bring all three raw materials together at this lime saturation. The two hypothetical mixes A and B can be blended to any silica ratio possible with the given compositions of the raw materials. If the value selected from experience of the reactivity of these raw materials is 3.0 and if the percentages of Mix A and Mix B needed are a and $100-a$, respectively, then:

$$\frac{S}{A+F} = \frac{13.09a + 16.05(100-a)}{(5.30+2.17)a + (1.42+0.67)(100-a)} = 3.0 \qquad (2.4)$$

This yields a value of 51.2% for a and the composition of the three mixes is therefore as follows.

Mix	Clay: %	Loam: %	Chalk: %
A	22.3		77.7
B		16.6	83.4
C	11.4	8.1	80.5

An analysis of a typical coal ash is included in Table 2.1 and the incorporation of this low lime material must be allowed for in proportioning the raw materials for a coal-fired kiln. If the ash content of a dry coal is 10% and 80% of it is absorbed in the kiln and if 25 t of the coal are needed to produce 100 t of clinker, then the ash incorporated is $25 \times 0.10 \times 0.80 = 2.0$ t.

The composition of the clinker from Mix C (i.e. after it has lost 36.1% of its weight as CO_2 and H_2O) is given in Table 2.1, and incorporation of the above quantity of ash gives clinker D. It is immediately apparent that there is a marked fall in lime saturation and so for a coal-fired kiln it is necessary to prepare a raw feed composition with an LSF high enough for the anticipated ash absorption. The figures also emphasise the need to use coal that is consistent in quality and to ensure the efficient, uniform absorption of the ash.

The method of calculation described has the advantage that when using a simple calculator, checks are possible after the analysis of each mix has been obtained. It also illustrates the principle that to fix x ratios, $x + 1$ materials of suitable composition are needed. It is also possible, of course, to find the composition of the final Mix C directly by solving a set of simultaneous equations derived by substituting the chosen values of LSF

Table 2.2. Effect of oxide content on potential compound content and liquid formation at clinkering temperatures

LSF	100%				90%	
A/F	1.5		3.0		1.5	
S/A + F	2.0	3.0	2.0	3.0	2.0	3.0
Analysis	%	%	%	%	%	%
SiO$_2$	20.88	22.42	20.70	22.28	22.42	24.10
Al$_2$O$_3$	6.27	4.49	7.77	5.57	6.72	4.82
Fe$_2$O$_3$	4.18	2.99	2.59	1.86	4.49	3.21
CaO	68.69	70.10	68.94	70.29	66.38	67.86
Compound content						
C$_3$S	72.8	80.4	67.4	76.7	48.1	55.9
C$_2$S	5.0	3.7	8.6	6.1	28.0	27.0
C$_3$A	9.5	6.9	16.2	11.7	10.2	7.3
C$_4$AF	12.7	9.1	7.9	5.7	13.7	9.8
Liquid formed at						
1338°	25.4	18.2	15.8	11.3	27.3	19.6
1400°	27.7	19.8	28.6	20.6	29.7	21.3
1450°	28.2	20.2	29.1	20.9	30.2	21.7

and S/R and the three chemical analyses into the expressions for these ratios. Such simultaneous equations are now usually solved by a computer, which can rapidly provide a set of mix compositions for a chosen set of values for each ratio and also indicate unachievable ratios as negative solutions. Another method of proportioning raw materials is to base the calculation on target values for the C$_3$S content of the clinker using equations such as those in Section 1.2.3. It can be seen in Table 2.2 that a high content of C$_3$S corresponds to a high LSF and a high silica ratio.

2.3 Reactivity of the raw materials

Reactivity is assessed as part of mix selection because a mix with ideal chemical composition may require too high a temperature to achieve adequate combination of lime. Only the rate at which the final equilibrium state is approached is of practical interest and this could be evaluated by the determination of unreacted lime in samples of clinker heated for different times at each of a series of fixed temperatures. However, the nature of the rotary kiln makes the use of a fixed time at a range of temperatures more appropriate. The complete laboratory evaluation of a set of raw materials involves the examination of the effect of variations in composition and fineness of the mix on its reactivity using a fixed heating

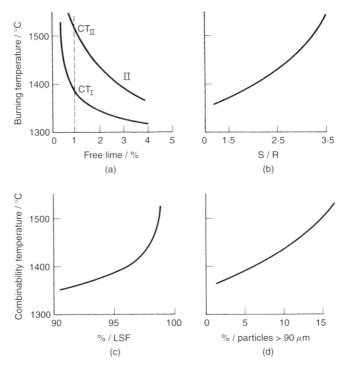

Fig. 2.3. Factors influencing the combination of cement raw materials: (a) combinability temperatures (CT) for two samples; (b), (c), (d) effects of silica ratio (S/R), lime saturation factor (LSF) and percentage of coarse particles on combinability temperature

regime. This is chosen to simulate kiln conditions and involves a period of 20–30 min at maximum temperature, precise conditions depending on an experimental comparison between practical and laboratory results.

Figure 2.3(a) shows the form of typical *combinability* or *burnability curves*. The form of such curves depends on:

(a) the intrinsic reactivity of the solids present at the clinkering temperature,
(b) the percentage and nature of the melt,
(c) the lime-saturation factor,
(d) the particle-size distribution of the solids.

Since the combination of only the last few percent of lime is relevant, the least reactive solids and their particle sizes are major influences on the form of the combinability curve. The dominant substances are lime, if it is present in clusters of crystals derived from coarse limestone particles, and free silica in the form of coarse quartz crystals. The lower degree of combination at all temperatures for Mix II in Fig. 2.3(a) could be due to the presence of a greater proportion of coarse particles of these substances.

Figure 2.3 gives typical results of experiments to identify an optimum mix with a given set of raw materials. From curves such as those in 2.3(a) a characteristic temperature is noted — *the combinability temperature* — which is that at which the free lime is reduced to some arbitrary level, such as 1%. This temperature is then plotted for a series of mixes with a range of compositions and finenesses as shown in Fig. 2.3(b), (c) and (d).

In Table 2.2 some results calculated by the Bogue method are given to illustrate the effect of variation in oxide content on clinker compound composition. The table also demonstrates the convenience of considering composition in terms of the commonly employed ratios, since these are more readily mentally linked with trends than the oxide analyses themselves. The compositions in Table 2.2 are for the pure quaternary system and in commercial clinker the simple Bogue potential compound contents might only total 95%, the rest being made up of the minor phases. Solid solution of some minor components and their contribution to liquid formation during clinkering produce further deviations from the results given. Although methods of allowing for these effects have been proposed (Taylor, 1984), the simple Bogue procedure is frequently employed.

The percentages of liquid in Table 2.2 were calculated using the equations derived by Lea and Parker in their study of the quaternary system in which liquid first forms at 1338°. Chemical formulae represent the mass percentage of an oxide.

Percentage liquid at:

$$1338°\quad 6.1Fe_2O_3$$
$$1440°\quad 2.95Al_2O_3 + 2.2Fe_2O_3$$
$$1450°\quad 3.0Al_2O_3 + 2.25Fe_2O_3$$

These equations apply to clinkers with an alumina to iron oxide ratio (A/F) greater than 1.38, which is usual for ordinary Portland cement. For ratios below 1.38 (for example, in sulfate-resisting cement — Section 9.2.2) the formula for 1338° is $8.2Al_2O_3 - 5.22Fe_2O_3$.

Points to note in Table 2.2 are:

(a) the proportion of silicates increases with increasing silica ratio while at the same time the proportion of liquid falls at all temperatures,

(b) the proportion of tricalcium silicate increases considerably with increase in lime saturation between 90 and 100%,

(c) the proportion of liquid at 1338° decreases with increasing A/F, at a fixed silica ratio.

Increasing A/F above 1.38 reduces clinkering efficiency in a kiln as material approaches the hottest zone (1400°–1450° commonly) because of the smaller amount of liquid formed at 1338°. For a given rate of transport of material through the kiln, the time during which it contains an appreciable quantity of melt is therefore reduced as this ratio increases. In addition, a high proportion of alumina in the flux makes it more viscous, which reduces the diffusion rates of ions in it, retarding the formation of C_3S (Section 3.4.3). Consequently, where the indigenous raw materials yield a mix with A/F as high as 3.5–4.0, importing iron oxide is considered desirable.

The objective of reactivity studies is to determine the best way to use the raw materials from both a clinker quality and an economic point of view, the latter involving both material preparation and burning costs. The strength-generating properties of a cement are enhanced by a high content of silicates and, at early stages of hydration, by a high C_3S content. To obtain these, a high silica ratio and a high LSF are needed but, as shown in Fig. 2.3, increasing these ratios increases the combinability temperature and a compromise is necessary. Problems attributable to particle size arise particularly if a hard limestone or quartz is present in the mix. If sand is added to increase the proportion of silicates then it is usually advantageous to grind it separately.

2.3.1   Model for combinability

The treatment of raw materials combination given in the last section involves direct experimental evaluation. A number of attempts have been made to identify and quantify the fundamental processes controlling the rate of the final stage of clinkering. The work of the Danish group, which developed that of Kondo and Choi (1962), will be considered here since it has been linked with combinability. Other fundamental work on clinkering kinetics will be considered in Chapter 3.

Christensen (1979) and Johansen (1979) considered a model mix in which coarse particles of calcite and quartz are involved in separate local 'equilibria' based, for example, on compositions I and II in Fig. 1.1. Composition I would be centred on a cluster of lime crystals (derived from calcite) which interacts at 1600° with liquid compositions on the liquidus between Z and the C–A axis. Coarse siliceous particles, on the other hand, initially react to form a cluster of C_2S crystals (Fig. 4.1) which, with localised composition II, interact at 1600° with liquid compositions above T on the liquidus. Regions of three types are relatively quickly established in the clinker: lime-centred regions separated by bands of C_3S particles from regions rich in C_2S. At 1600° the liquid compositions at the CaO–C_3S and C_3S–C_2S boundaries are Z and T, respectively.

The difference in lime concentration between the two boundary liquids (ΔC) provides the driving force for the growth of a C_3S region at the expense of a C_2S region, predominantly by the dissolution and diffusion of lime. Christensen and Johansen (1980) gave an equation (2.5) relating the increase in the width of the C_3S region (Δx) with time (t) at the burning temperature:

$$\Delta x = \sqrt{(kt)} = \sqrt{(2D\alpha\Delta C\,H)} \qquad (2.5)$$

where D is the coefficient for lime–silica interdiffusion, α the melt content in the C_3S region, and H is a function of the difference in lime contents between local compositions I and II and those on the C_3S primary phase field boundaries on the straight line joining I and II. This last term is intended to allow for microheterogeneities. The parabolic law has been confirmed experimentally using diffusion couples (Johansen, 1989).

In practice there is a limit to the size of non-equilibrium region which can be eliminated by diffusion during the clinkering in a cement kiln. The Danish workers chose 30 minutes at 1400° as reference conditions for establishing burnability and 125 μm for calcite and 44 μm for quartz as realistic critical particle (sieve) sizes. Fundal (1979) derived empirical equations relating residual-free lime (C_{1400}), LSF of the mix, silica ratio (modulus) M_s and the percentages of coarse particles in the mix of the form:

$$C_{1400} = 0.33[\%LSF - \%LSF(M_s)]$$
$$+ 0.93[\%SiO_2(+44)] + 0.56[\%CaCO_3(+125)] \qquad (2.6)$$

where

$$\%LSF(M_s) = 107 - 5.1M_s \text{ for } 2 < M_s < 6 \text{ and } 88 < LSF < 100 \quad (2.7)$$

The first term in Equation (2.6), the so-called chemical contribution allowing for random microheterogeneities in the mix, is determined experimentally and Fundal's result obtained by correlation analysis of results for 24 mixes is given in Equation (2.7). Equation (2.6) can be used to predict the effect on residual free lime of coarse particle content for mixes containing the same minerals as rate-determining coarse particles. Fundal gave a set of constants for use if other or additional minerals take on this role and has more recently refined his model (1996).

Empirical equations can also be derived to relate changes in residual free lime after a chosen period to changes in burning temperature. These make it possible, for mixes of similar chemistry, to deduce the combinability curve from a determination of free lime at one temperature only. This implies that the combinability curve is displaced as temperature changes but its shape remains unaltered. Such equations are useful when the combinabilities of a large number of similar mixes need to be determined.

2.4   Physical properties of raw materials

The ease of preparation of raw materials for blending and burning is principally determined by their crushing strengths and abrasivities, as well as their moisture contents. Strengths cover a wide range — for example, from $75\,N/mm^2$ (chalks) to over $300\,N/mm^2$ (hard limestones) — because of differences in both grain size and bonding forces. No material except silica, however, is comparable to cement clinker in difficulty of grinding. This is related to the strength and hardness of these materials, the latter property being related directly to ability to cause abrasive wear. The abrasivity of a raw mix is assessed by shearing a dried, crushed sample in what resembles a concentric cylinders viscometer. Small, removable inserts of the appropriate alloy are set into the outer cylinder and weighed at intervals to determine wear rate.

Because of the dependence of the size-reduction process on a number of variables, and also because there is no adequate theory of size-reduction equipment, forecasts of the power requirements of grinding mills are based on empirical laboratory tests which aim to simulate full-scale operation. Several such tests are commonly applied in parallel to increase the reliability of estimates.

2.4.1   Grindability

The power needed to grind the raw materials used in making clinker, or in grinding the latter with gypsum, can be deduced from measurements made with a laboratory or pilot-scale mill. Cement makers and cement plant manufacturers usually employ procedures which they have developed independently. The electric power used in driving the mill and its charge of material (including steel media in a ball mill) is experimentally related to the increase in surface area produced. For a ball mill, the correlation with continuous milling is obtained by periodically separating fine material by sieving and replacing it with fresh material and also reducing the media size as grinding proceeds. The sizing of a vertical spindle (roller) mill (Fig. 3.4) is based on pilot-scale trials involving 1–2 t of material. Although energy efficiency makes this mill the preferred choice, some raw materials are too abrasive and a ball mill is employed. Specially hardened alloys are used for the media in a ball mill and, to maintain grinding efficiency, regular inspection and replacement of worn media are essential.

Materials may be ranked on the basis of the curve of surface area produced as a function of power input on a *grindability* scale, the number assigned increasing with increasing difficulty of grinding. In this approach, a limestone with a grindability of 110 requires 10% more energy for grinding to a given fineness than the standard, taken as 100. Cement milling and characterisation of the product are discussed in Chapter 5.

3. Production of cement clinker

3.1 Introduction

In Chapter 2, the objective in making cement clinker, namely the combination of the four principal oxides to make a material high in di- and tri-calcium silicates but low in free lime, was discussed in terms of the composition and reactivity of the raw materials. In this chapter, commercial methods used will be outlined and some of the chemical and physical processes that occur at high temperature considered. The manufacturing process is primarily concerned with the selection of the most efficient engineering methods for crushing, grinding, blending and conveying of solids on a large scale as well as with their heat treatment (*pyroprocessing*). Energy usage is considerable and is constantly monitored so that improvements can be made.

Two distinct processes are employed in the production of clinker. In the *wet process* a slurry of the finely divided raw materials is made and pumped into a long rotary kiln. In the *dry process* the raw materials are prepared for pyroprocessing as a blend of finely ground powders and initial heating is usually carried out in a preheater using the hot gases from a relatively short kiln. However, in those parts of the world where the raw materials are relatively dry, and fuel costs are not prohibitive, drying, calcining and clinkering may be carried out in a long dry kiln. The wet process was at one time predominant but a rapid increase in fuel costs in the 1970s accelerated the final stages of its replacement, since it involves the evaporation of a substantial quantity of water, typically 30–35% of the mass of the kiln feed. However, in areas where the primary raw material is a porous, high moisture content chalk containing flint, the wet process has survived in the preparation of raw materials.

3.2 Preparation of kiln feed

After extraction from the quarry, raw materials must be crushed, ground and blended to make the *raw feed* (*meal*) for the kiln. The choice of equipment is dependent on the physical properties of the materials

(Section 2.4). The importance of adequate grinding and blending of raw materials cannot be overemphasised, as there is a limit to the size of the regions of different chemical composition which can be eliminated by the rate-determining processes of dissolution and ionic diffusion in clinkering (Section 3.4.3).

3.2.1   Wet and semi-wet processes

Soft materials are converted to a slurry with water in a *wash mill*. This involves vigorous agitation with harrows and tynes hanging into a cylindrical tank from a centrally pivoted rotating arm. Fine material in suspension passes through a vertical screen at the side of the tank, against which it is thrown by the harrows. The slurry produced, controlled by measurement of its density and the addition rates of the materials, should contain the highest concentration of solids at which it is pumpable. Reduction of the water content necessary for pumping can often be achieved by adding a deflocculant, such as sodium carbonate or silicate, at a cost which is less than that of the kiln fuel saved. However, if significant amounts of a smectite (swelling clay) are present in the clay, the addition required will be excessive.

Either clay or chalk may be slurried first and the second component blended with it in a second wash mill. When chalk contains flints it may be added to the clay slurry in a form of tube mill, called a wash drum, in which the flints act as grinding media and are then 'scalped off'. Remaining coarse material is removed from the slurry by fine screens or by hydrocyclones. The suspension enters a cyclone tangentially in an upper cylindrical section that produces a downward spiral motion in it (a vortex) which carries it into the conical section. The quantity of material leaving the bottom is limited so that an upward vortex is produced in the centre. Heavy material is thrown outwards and downwards and light material is carried up by the fluid flow to the second exit.

Further adjustment, using minor components such as ground sand, pulverised fuel ash or iron oxide, may be necessary to optimise chemical composition. The refined slurry is monitored by determination of residues when a sample is washed through standard sieves. Acceptable residues depend on the reactivity of the coarse material (Section 2.3.1) but 0.5% greater than $300\,\mu$m and 12% greater than $90\,\mu$m would be typical. Chemical composition is checked by sampling at several stages. An automated X-ray fluoresence spectrometer is used to determine silicon, aluminium, iron and calcium, and an on-line instrument can control equipment feeding the blending mill. The slurry is then held in tanks in which it is agitated both mechanically and by compressed air to prevent segregation.

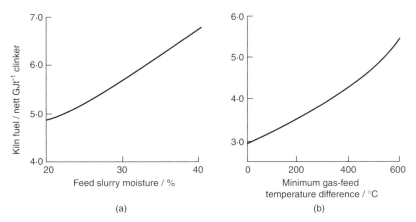

Fig. 3.1.   Factors primarily influencing fuel consumption in (a) wet process and (b) dry process (Davis, Stringer and Watson, 1979)

Fig. 3.2.   Filter press used in a semi-wet process plant (courtesy of Blue Circle Industries)

Water has a high enthalpy of evaporation, and so the curve related kiln fuel consumption to the water content of a slurry is a steep one (Fig. 3.1(*a*)). Consequently, a number of the remaining wet process plants have been converted to the *semi-wet process* by installing filter presses (Fig.

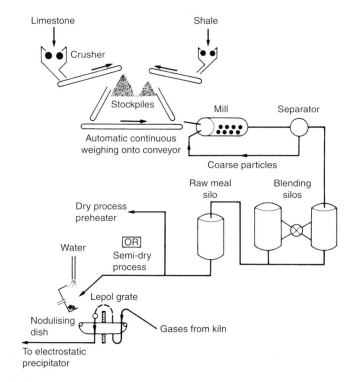

Fig. 3.3.   Preparation of raw meal for dry and semi-dry processes

3.2) which produce a cake with a water content of about 18–20% as feed for a kiln or preheater. In some variants of this process part of the kiln feed is introduced to the preheater in a dry state.

3.2.2   Dry and semi-dry processes

The sequences of operations used in preparing the raw meal for the dry and semi-dry processes are shown schematically in Fig. 3.3. Raw materials are crushed and put into stockpiles, usually under cover. A considerable degree of homogenisation is obtained by laying them down in strips or layers and systematically reclaiming from the stockpiles produced. They are then continuously proportioned into the milling and drying system by weight. Gases from the kiln or the cooler are used for drying, although supplementary firing may be necessary. Grinding is carried out either in a ball mill or a vertical spindle mill in which rollers grind the material in a pan (Fig. 3.4). The latter is now usually selected for grinding raw materials in a new plant (unless they are exceptionally abrasive) as it uses significantly less energy for a given fineness than a

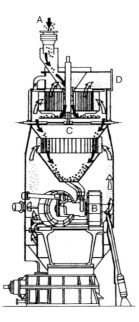

Fig. 3.4. Vertical roller (spindle) mill used in drying and grinding raw materials: A feed; B roller on grinding table; hot air flow (arrows) carries material to classifier — C — from which the finer particles are carried to exit — D. A peripheral dam ring determines the depth of the bed on the table which rotates. Hydrostatic pressure is applied to the rollers (courtesy of FL Smidth-Fuller)

ball mill. Early problems of wear and maintenance have been reduced by using replaceable wear-resistant alloy surfaces, and steadier running is achieved by recycling some of the ground product to establish a denser bed in the pan (Ellerbrock and Mathiak, 1994).

Separation of coarse from fine material leaving the mill may be effected by entraining the powder in exhaust gas from the kiln, followed by separation of coarse material in a *separator* and/or cyclone system. Alternatively, the mill product may be transferred to the separator by a bucket elevator. The separator or classifier makes use of the balance between centrifugal and air drag forces, produced by a rotating plate which strikes the particles and by a fan. The milled raw meal is transferred to blending silos, which may be more than 30 m high, and then to final stage silos. To reduce the capital investment required in the blending and storage silo system, some new plants now employ limited storage capacity with direct on-line chemical analysis and continuous adjustment of the kiln feed composition, using corrective materials such as ground limestone, sand and iron oxide.

3.3 Pyroprocessing: principal manufacturing processes

3.3.1 Introduction

The processes that occur when a raw meal is heated to clinkering temperatures, and where they take place in the principal types of plant employed, are summarised in Table 3.1. In *wet process* and some *semi-wet* plants they all take place in the rotary kiln, which can be divided into the zones in which they occur (Fig. 3.5). In the most modern *dry process* plants almost all but the clinkering is carried out by suspending the powder feed in hot combustion gases before it enters a short kiln.

Table 3.1.   Principal plant types used in pyroprocessing

Chemical/physical process	Drying	Decomposition		Melting/clinkering	Cooling
		Clay min.	Calcite		
Plant type		300–650°	800–950°	1250–1500°	
Wet/semi-wet	Drying*	Preheating*	Calcining*	⎫	
Semi-dry (Lepol)	Grate	Grate	Grate/kiln[†]	⎬ Kiln burning	Kiln/cooler
Dry	Raw mill	Preheater	Kiln	⎭ zone	
Precalciner	Raw mill	Preheater	Calciner/kiln[†]	⎭	

* Zones in the kiln. In the semi-wet process, cake may be dried on a grate or in a crusher/dryer.
[†] Minor amount in kiln.

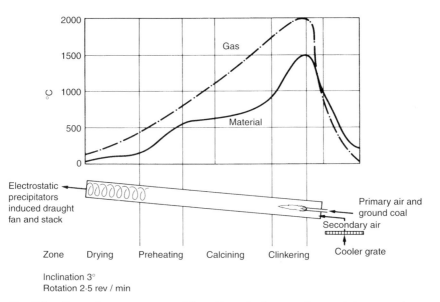

Fig. 3.5.   Zones in a wet process kiln with typical gas and material temperature profiles

Material moves down a rotary kiln by sliding and rolling induced by the rotation and inclination of the kiln. The powder begins to form nodules as melting begins in the approach to the clinkering zone, the driving force for this being the high surface tension and low viscosity of the melts formed. Timashev (1980) reported a linear dependence of nodule size on melt surface tension. Gas velocities in a kiln are high. In the wet kiln, where a high fuel input is needed for drying the slurry, velocities may exceed 5 m/s so that dust entrainment occurs. Any dust not captured by the slurry in the drying zone and leaving the kiln constitutes a significant energy loss since its 'sensible heat' (heat content) is lost. However, in the dry process most the dust/heat is retained in the preheater. Dust leaving the preheater may be captured where exhaust gases are used in a raw mill/drying system. Before going to atmosphere, gases are passed through a cooler and bag filter or a multi-chamber electrostatic dust precipitator after conditioning to a suitable humidity and temperature. Collected dust which is not too rich in alkali may be returned to the raw feed blending system.

The choice of process for a plant is determined by balancing fuel economy against the capital investment required and its depreciation, as well as the amount and quality of the cement demanded by the (mainly) local market. Where there is limited demand and a relatively unsophisticated product suffices, a simple vertical shaft kiln may be used. A new dry process plant is unlikely to be constructed to produce less than about 800 000 t per annum. Kilns are fired by pulverised coal, heavy fuel oil or natural gas, according to local availability and cost. Coal has usually been employed in the UK but with a substantial supplement of petroleum coke. Other low cost materials employed as fuel supplements include old tyres and solvent wastes, use of the latter being subject to the constraints imposed by any deleterious effect on cement quality, refractory life or potential emissions to atmosphere from the system.

3.3.2   Wet and semi-wet processes

A long kiln (length/diameter (L/D) ~ 30) is necessary to carry out the endothermic processes of drying the slurry, decomposing the clay minerals and the calcite, and then to raise the temperature of the feed to a level at which clinkering takes place. Some 30 minutes' residence in the burning zone is usually required. The strongly endothermic nature of both water evaporation and calcination is reflected in the plateaux in the material temperature profile in Fig. 3.5. Their length is an indication of the time taken for completion of these processes, involving minutes rather than the seconds needed in a fluidised powder suspension. In the drying zone, heat transfer from gas to slurry is optimised by heavy chains fixed to the shell of the kiln. These transfer heat from the combustion gases to the slurry and lift it to increase the surface at which evaporation can occur. In the semi-wet process, filter cake is either introduced to a shorter kiln chain section or it enters a kiln after being dried on a Lepol grate or in a crusher-dryer.

3.3.3 Dry processes

It has long been recognised that heat transfer to a powder is much more efficient when it is suspended in a hot combustion gas, although a fluidised bed treatment has not yet been efficiently applied to a process such as clinkering in which partial melting is involved (Erhard and Scheuer, 1994). In a modern *dry process* plant, the blended raw materials are first heated by the hot gases from a kiln in a *suspension preheater* before entering the kiln. In the process now used in most new large scale plants, part of the fuel is burnt in a furnace (*precalciner*) integrated into the preheater, so that feed enters a short kiln with a greater part of the calcination completed.

In a suspension preheater, gases leaving a relatively short kiln ($L/D \sim$ 10–15) at a temperature in excess of 1000° raise the raw material temperature to about 800° in a system of heat exchangers arranged vertically. Several designs exist and one system is illustrated schematically in Figure 3.6. The raw meal is heated in co-current, separated from the gas stream

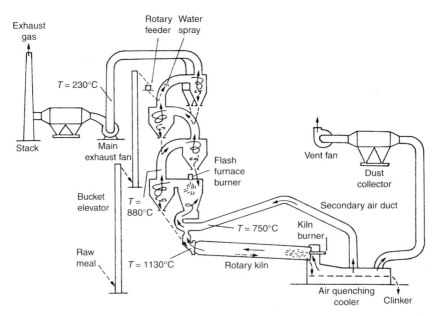

Fig. 3.6. Dry process with precalciner (typical gas temperatures indicated) (courtesy of FL Smidth-Fuller)

by a cyclone and then dropped into the riser pipe of the next stage. More than one vertical string of cyclones may be used to increase throughput. In other systems, such as the Krupp, counter-current flow of powder and gas is maintained in vortex chambers.

In the most modern high capacity plants, a furnace (precalciner) is incorporated at the base of the preheater in which up to 60% of the total fuel is burnt, thus avoiding most of the much slower and less efficient decomposition of calcite in the kiln. Several different arrangements of the plant components are in use (Erhard and Scheuer, 1994). Where the clinker leaving the kiln is cooled on a grate, the excess of exhaust air over that required for combustion in the kiln is used to burn the fuel introduced into the calciner. The major advantage of this modification to the dry process is an increase in clinker output for a given capital investment. The high CO_2 partial pressure generated by the combustion of the fuel and the decarbonation of the calcite in the raw feed prevents completion of the calcination. However, this has been achieved in a two-stage calciner where the partial pressure of CO_2 is much lower in the second stage.

After calcination the feed enters a rotary kiln at a temperature of around 1000°. Kilns with an L/D as low as 10 have been used, but the reactivity of the components of the raw feed must be taken into account in determining the time/temperature regime needed to complete clinkering. Thermal efficiency in a dry process with a suspension preheater can be expressed as a function of the relationship between kiln fuel consumption and the minimum gas-feed temperature difference achieved in the preheater. Clearly, the lower the temperature of the gases leaving the system, the greater the degree of heat exchange achieved and the smaller the amount of kiln fuel needed (Fig. 3.1(*b*)). However, if the exhaust gases do not have sufficient drying power for the raw milling circuit, supplementary fuel will be needed. In some modern plants in Japan, for example, a higher exhaust gas temperature is maintained for use in power generation. In any process, the exhaust gas must leave the system with a temperature–humidity relationship such that it clears the stack without reaching its dew point.

3.3.4   Semi-dry (Lepol) process
This process dries and partly calcines the nodules produced in feed preparation on a moving grate using gases from the kiln. It was developed in parallel with suspension preheater systems but avoided the early problems which these raised with dedusting the gases leaving the preheater and passing to atmosphere. The development of efficient and reliable exhaust gas cleaning systems has rendered it obsolescent.

Raw meal is nodulised in a rotating, inclined shallow dish which is fed with a stream of the powder and a fine water spray (Fig. 3.3). Solid particles agglomerate on the surface of the water droplets and the agglomerates increase in size and number, eventually spilling over the rim of the dish. The driving force for the agglomeration is the surface tension of the water. The diameter of the nodules required, usually about 15 mm, is controlled by the dimensions of the dish, the rate of its rotation and the rate at which powder and water are fed to it.

The nodules, which have a water content of about 12%, are dried and then largely decarbonated on a moving *Lepol grate* by gas from the kiln which passes downwards through a bed of them in each of two chambers (calciner and then dryer). The efficiency of this process depends on the strength and abrasion resistance of the nodules, especially where they fall from the grate into the kiln. At this temperature the shale has lost its original plasticity but nevertheless this property can be correlated with nodule performance. If the nodules are too friable then enough dust may be blown back from the kiln to restrict the passage of hot gases through the nodule bed (*blinding*).

3.3.5   Clinker cooling

Once material has passed the kiln flame, its temperature begins to fall and it drops into a cooler, leaving the kiln with a surface temperature of about 1100°. The clinker cooler has two functions, to reduce the temperature to a level at which the clinker can be transported to storage and cement milling and to recover as much heat as possible in the form of preheated air for fuel combustion. There are two main types of cooler: *grate* and *integral*. The former consists of a moving metal grate through which air is blown by fans and, since air passes only once through the clinker, more air is required for cooling than is necessary for kiln fuel combustion and some heat is lost in exhaust gas. The integral or planetary cooler consists of a set of tubes disposed around the kiln in which counterflow heat exchange takes place with incoming air. Cooling is aided by metal lifter plates which increase the tumbling of the clinker. Capacity is limited and additional external cooling of the tubes may be necessary.

3.3.6   Refractories

An important element in the successful operation of a cement kiln is the selection and installation of the refractories lining it. They have to withstand abrasion as well as thermal and mechanical stresses that change as the kiln is rotated and, in the burning zone, chemical attack by the clinker as well as very high radiant temperatures when not shielded by the clinker bed. The conditions experienced by the refractories lining preheaters, Lepol grates and coolers are far less severe.

Where there is no chemical interaction with the kiln feed, grades of aluminous firebrick (25–80% Al_2O_3) are used according to the prevailing gas temperature. In the burning zone where the aggressive clinker melt is present, dense magnesite, dolomite or magnesia spinel bricks are used. Abrasion-resistant castables are used at the inlet and exit of a kiln and also in preheaters. Bricks with a dense abrasion-resistant working face and a more porous (lower thermal conductivity) backing have been developed for all but the hottest zones of a kiln to reduce heat loss.

Dimensional standardisation and careful installation of bricks, which are fitted in rings, as well as slow initial warm-up are essential. When the kiln is running, good control of the raw meal composition and feed rate and a steady flame all increase brick life. Monitoring of the steel shell of the kiln is necessary to detect any refractory failure. If distortion of the kiln shell occurs as a result of overheating, the refractories are subjected to additional severe stresses. Refractory replacement involves high material and other costs plus significant losses in production. The conditions experienced by refractories in a cement kiln are not provided in standard laboratory tests but refractoriness under load (BS 1902) is sometimes taken to indicate suitability for particular situations in a kiln.

Some interaction between clinker and refractory occurs in the burning zone and the coating produced is essential to limit the temperature of the kiln shell. It also reduces abrasion of the bricks as well as the variation in refractory temperature as the kiln rotates. Excessive build-up of coating can occur at the approach to the burning zone, forming *clinker rings* which inhibit material flow. Their formation is considered to be related to the composition of the raw materials and the temperature profile in the kiln, involving some melting and the transient formation of intermediate compounds.

3.4 Pyroprocessing: physical and chemical processes involved

This section considers the physical and chemical basis of processes occurring as the temperature of a raw meal is raised to the maximum, reached as it passes the flame.

3.4.1 Preheating

The decomposition of the clay minerals takes place in the temperature range 350–650°. The almost amorphous intermediates formed initially by dehydroxylation do not subsequently recrystallise because they react with calcite before high enough temperatures are reached. Any magnesite present will also decompose in this part of the system, its equilibrium decomposition pressure, $P^e_{CO_2}$, reaching 1 atmosphere at about 400°.

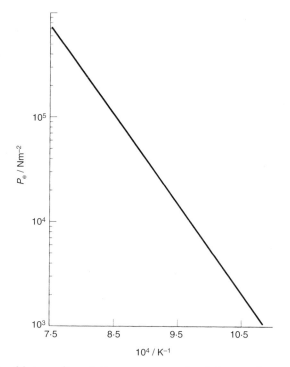

Fig. 3.7. Equilibrium dissociation pressure of calcium carbonate (P_e) as a function of reciprocal temperature (in Kelvin). The slope of the plot is $-\Delta H / R$ (Equation (3.1))

3.4.2 Calcining

The decarbonation of calcite is strongly endothermic. Specific rather than molar enthalpies are usually quoted because they are used in calculating the theoretical energy needed to produce 1 kg of clinker. Values of the equilibrium partial pressure of calcite have been determined as a function of temperature by a number of workers, with significant discrepancies. Fig. 3.7 is based on mean values given by Seidel *et al.* (1980). The relationship between the equilibrium constant K_P for decarbonation ($= P^e_{CO_2}$) and temperature T (in Kelvin) is given by the Van't Hoff equation which, in the integrated form, is:

$$\ln \frac{(K_p)_2}{(K_p)_1} = \frac{-\Delta H}{R} \left(\frac{1}{T_2} - \frac{1}{T_1} \right) \tag{3.1}$$

where ΔH is the molar enthalpy for the reaction and R is the gas constant. Equation (3.1) is only applicable over a range of temperature in which ΔH is sensibly constant to within the error in equilibrium pressure

Table 3.2.   Enthalpy changes for kiln processes

Process	Specific enthalpy (kJ/kg of substance starred)
Evaporation of water*	+2452[†]
Dehydration of clay minerals (referred to gaseous water)	
Kaolinite*	+538
Pyrophyllite*	+224
Decarbonation of $CaCO_3$ $20°$	+1765[†]
$890°$	+1644[†]
$2C + S \rightarrow \beta C_2S$*	−734
$3C + S \rightarrow C_3S$*	−495
$3C + A \rightarrow C_3A$*	−27
$4C + A + F \rightarrow C_4AF$*	−105 ($20°$)

[†] Data from Weber (1963); all other values courtesy of H.F.W. Taylor.
Values are for reactants starting and products finishing at $25°$ except where another temperature is specified.
For the purposes of the overall application of Hess's Law to the calculation of heat requirement for clinkering, the standard heats of formation of α-alumina and α-iron oxide can be used. This treats them as hypothetical decomposition products of the clay minerals which subsequently react to form the final compounds.

measurements. Over large temperature intervals, ΔH changes significantly because the difference between the specific heats of reactants and products changes (Table 3.2).

The factors determining the overall rate of decarbonation are:

(*a*) the material temperature which determines $P^e_{CO_2}$;
(*b*) the gas temperature which determines heat transfer rate;
(*c*) the external partial pressure of CO_2, which is the sum of the partial pressure derived from the calcite decomposed and from the combustion of the fuel;
(*d*) the size and purity of the calcite particles.

In the wet or long dry kilns the strongly endothermic nature of the reaction and the depth of the bed (although agitated) result in heat transfer being rate controlling.

If the individual calcite particles are suspended in the combustion gas stream then possible rate-controlling factors are:

(*a*) the rate of heat transfer through any already decarbonated material to the undecomposed core of a particle (Fig. 3.8);
(*b*) the reversible process of breaking a chemical bond in the carbonate ion and rearrangement of ions to form CaO crystals:

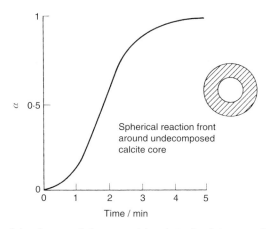

Fig. 3.8.   Plot of the degree of decomposition (α) of calcite as a function of time at 850° and the 'shrinking sphere' model

$$(Ca^{2+} + CO_3^{2-})_s = (Ca^{2+} + O^=)_s + CO_2 \tag{3.2}$$

(*c*) the diffusion of CO_2 through the product layer of lime crystals.

The slowest of these steps will determine the overall rate of decarbonation. For large crystals, high gas temperatures and low CO_2 partial pressures, heat transfer will be the rate limiting factor, while at lower temperatures (when rates of heat absorption needed to maintain equilibrium are low) with high external CO_2 pressures, diffusion of CO_2 will be rate controlling.

A plot of the degree of decarbonation (α), of calcite, measured by weight loss, against time (*t*) at constant furnace temperature is sigmoidal (Fig. 3.8). After the acceleration or induction period, which is very short at high temperatures, α and *t* are related by Equation (3.3) in which *k* is a rate constant:

$$(1 - \alpha)^{1/3} = kt \tag{3.3}$$

Such an equation may be theoretically derived by assuming that calcite particles are spherical and that the rate of decarbonation is proportional at any instant to the interfacial area between carbonate and lime.

Hills (1968) examined the rate of decarbonation of polycrystalline spheres (10 and 20 mm diameter) of calcite previously sintered in an atmosphere of CO_2. Sectioning established that the interface between lime crystals and undecomposed calcite advanced during decarbonation in a uniform manner towards the centre of the sphere. However, Hills showed that rate curves fitting Equation (3.3) do not distinguish between chemical reaction, heat transfer and CO_2 diffusion processes as rate-controlling. He found that the temperature inside the calcite spheres dropped to as much as 45° below that of the furnace when decarbonation was started by changing the atmosphere from CO_2 to air. He concluded that heat transfer and CO_2 diffusion (mass transfer) were the rate-controlling factors and derived an equation showing that the temperature dependence of the rate of decarbonation in the range 700–900° is, under most practical conditions, almost entirely determined by the temperature dependence of the equilibrium partial pressure of CO_2. Since the latter is determined by the enthalpy of the reaction (Fig. 3.7) the apparent activation energy (E_A) derived from a plot of log (reaction rate constant) against $1/T$ is ΔH (167 kJ/mol). E_A values of this magnitude had previously been found but taken to indicate that calcite decomposition is controlled by a chemical step since the temperature coefficient of simple diffusion processes is usually about 20 kJ/mol.

Khraisha and Dugwell (1992) determined rates of calcination of a natural limestone with particle sizes in the range 40–90 μm using a reactor simulating the conditions in a commercial precalciner. They monitored exit gas composition with an on-line mass spectrometer. Measurements of calcination rate were made over the temperature range 1076–1156 K in air, and in a mixture of 15% CO_2 with O_2 and N_2. The results were in accord with the shrinking sphere model and suggested that, under the conditions employed, chemical reaction was rate-controlling with an activation energy of 196 kJ/mol. Simultaneous injection of coal and limestone into the reactor was found to enhance combustion of the coal but inhibit calcination by increasing the partial pressure of CO_2.

Direct chemical interactions between calcite and the decomposition products of clay minerals also occur in this zone of the kiln, their importance depending on the fineness and intermixing of the original minerals. These endothermic reactions, which occur at slightly lower temperatures than normal calcite decomposition, are detected by differential thermal analysis as a shoulder on the low temperature side of the decarbonation endotherm. Solid–solid reactions increase in extent throughout the calcination zone and fluorides, sulfates, and alkalis may produce melts which, although small in amount, promote reaction. At the contact between two reacting particles, such as lime and silica, a range of compositions would be expected to develop, corresponding to the sub-solidus of the binary system at the temperature involved. C_2S is formed from about 800° upwards. Some product phases are likely to be formed in very small bands, only revealed by determination of atomic ratios using X-ray microanalysis (Section 4.5). Intermediate products which have been found in amounts revealed by conventional X-ray diffraction, in samples taken through ports in a kiln shell, are $2C_2S.CaCO_3$ (spurrite), $C_{12}A_7$ and C_2F.

3.4.3   Clinkering (sintering in the presence of a liquid phase)

After the completion of decarbonation, material temperature rises rapidly and all but the coarsest siliceous material is quickly converted to C_2S. Acidic melts form in silica-rich regions where interactions with lime, alumina and/or iron oxide can occur. For example, melting occurs from $1170°$ in the C–A–S system. In the first stage of clinkering the melt formed exerts significant capillary forces as a result of its high surface tension, producing rapid shrinkage and nodule formation if this has not already occurred. The constitution of the material soon approaches that described in Section 2.3.1, that is high and low lime regions separated by a developing zone of C_3S. Both Russian and Japanese researchers have shown that in a slower, second stage of clinkering the fractional shrinkage $\Delta L/L_0$ with time, t, can be fitted to an equation proposed by Kingery (1959) for the sintering of a solid in the presence of a reactive liquid:

$$\frac{\Delta L}{L_0} = \left(\frac{6k_2\delta\, DC_0\, \gamma V_0}{k_1 RT} \right)^{\frac{1}{3}} r^{-\frac{4}{3}} t^{\frac{1}{3}} \tag{3.4}$$

where k_1, k_2 are constants; δ is the thickness of the liquid film between the particles; D is the diffusion coefficient in the liquid; C_0 is the solubility of the solid, V_0 its molar volume and r its initial particle size, γ is the liquid vapour interfacial tension.

The processes finally bringing the clinker closer to equilibrium involve the dissolution of lime and C_2S in the melt, diffusion of ions and crystallisation of C_3S. Identification of the rate-controlling step or steps therefore requires independent measurement of these processes. Much of the available data on the viscosity and surface tension of clinker melts and on diffusion in them was published by Butt, Timashev and their co-workers and reviewed by Timashev (1980) and Seidel et al. (1980). Viscosities were measured with a rotational viscometer for two melt compositions; the $1338°$ eutectic and an equilibrium melt formed at $1450°$ (S:A:F:C were 6.0:22.7:16.5:54.8 and 7.5:22.6:12.9:57.0, respectively). Since these liquids are extremely aggressive, platinum equipment is necessary. Surface tension was measured by the bubble pressure method and radioisotopes were used to obtain diffusion coefficients. By measuring these as functions of temperature, Arrhenius activation energies for diffusion were derived (Table 3.3). The most rapidly diffusing ion, Ca^{2+}, determines the rate of C_3S formation and this will therefore occur by extension into the C_2S zone. Undissolved C_2S is frequently encapsulated in the process. Silicon exists as discrete SiO_4^{4-} ions in these melts, diffusing by rotation and place exchange with other anions. Aluminium and iron exist as M^{3+} and MO_4^{5-} ions. Calcium is present as simple cations and where the diffusion of lime is referred to,

Table 3.3.   Diffusion in a clinker melt (Butt, 1974)

	Diffusion coefficient at 1450°: m²/s	Activation energy (1450–1525°): kJ/mol
Ca	5.3×10^{-9}	164
Fe	5.7×10^{-10}	294
Al	2.4×10^{-10}	335
Si	4.7×10^{-11}	368

Melt composition (weight %): S — 7.5; A — 22.6; F — 12.9; C — 57.0

the overall compositional balance in diffusion, ignoring its mechanism, is meant.

The activation energy for viscous flow was found to be 364 kJ/mol and the similarity to E_ASi indicates the dominance of the silicate ion on melt viscosity. Rates of solution of CaO and C_2S increase with increasing temperature and, because dissolution involves diffusion of ions away from the liquid immediately adjacent to the solid surface, they also increase with decreasing melt viscosity at constant temperature. Results showing some of the relationships found are given in Fig. 3.9. Those minor constituents in raw materials, such as MgO and SO_3 (added as $CaSO_4$ but to give the same total CaO in the resulting melt) which reduce melt viscosity, were shown to increase diffusion rates but alkalis had the reverse effect. The influence of alkali and sulphate ions added simultaneously was found to be complicated by their forming immiscible liquids (Timashev used the term 'liquation' for this).

In the rotary kiln three additional processes complicate the simple dissolution–diffusion–crystallisation model. These are:

(a) movement of material involving shear and compaction within nodules, which aids chemical combination;

(b) volatilisation of alkalis and sulfur oxides;

(c) the existence of low partial pressures of oxygen with some reduction of Fe^{3+} to Fe^{2+}. Reduction, if excessive, may destabilise C_3S and hence adversely influence cement quality (Locher, 1980), but is minimised by proper flame control.

The extent to which equilibrium is approached in clinkering reactions is determined by the several rate processes described in this section, but the limiting factor in the final stages is often the rate of dissolution in the melt of the least reactive particles in the $> 90\,\mu m$ size fraction of the kiln feed (Section 2.3.1). Coarse quartz particles, for example, produce relatively large silica-rich regions in a kiln feed. The quartz usually reacts completely with lime to form clusters of belite crystals but conversion of the belite to alite may not be completed in the time for which clinker is in

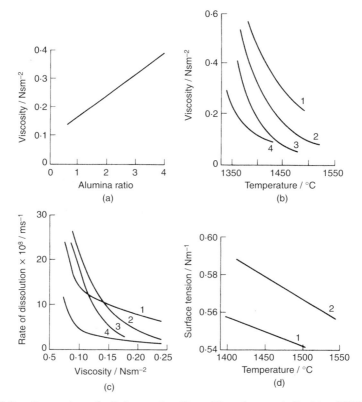

Fig. 3.9.   Properties of clinker melts (Butt, Timashev and Osokin, 1976): (a) effect of alumina ratio on the viscosity of 1440° melts (Budnikov, Entin and Belov, 1967); (b) effects of temperature and additions on the viscosity of a 1450° melt: (1) 3% K_2O, (2) no addition, (3) 2.5% SO_3, (4) eutectic melt; (c) relationship between changes in melt viscosity (effected by changing temperature) and the dissolution rates of CaO and C_2S: (1) CaO in eutectic melt, (2) CaO in 1450° melt, (3) C_2S in 1450° melt, (4) C_2S in eutectic melt; (d) effect of temperature on the surface tension of (1) eutectic melt and (2) 1450° melt. For composition of the melts in (b), (c) and (d) see text

the burning zone. An examination of the interdependence of kiln feed 90 μm residue and clinker quality (strength-generating potential) at 15 cement manufacturing plants was reported by Moir (1997).

Potassium and sodium sulfates and chlorides are volatile at kiln burning zone temperatures. At 1400° the vapour pressures of the chlorides are more than 1 atm, while those of the sulfates are approximately 0.5 and 0.2 atm for sodium and potassium respectively. A cycle is set up involving evaporation of these minor constituents of the raw materials in the burning zone and their return to the kiln after condensation on the cooler material entering the system. Condensation of sulfates on the cooler walls of the system can lead to a build up of raw feed adhering to the deposits and a serious restriction in material flow. The levels entering the clinkering zone increase until a steady state is reached, in which there is a balance between what is entering this zone and what is leaving it by evaporation and in the clinker.

It is necessary to maintain a small but sufficient excess of oxygen in the kiln burning zone to avoid reduction of sulfates. This significantly increases SO_2 recycling between the burning zone and cooler parts of the system, where it is absorbed by incoming calcareous material. If the raw materials contain chloride, even at levels as low as 0.02%, or if a low alkali cement is required, then some thermal efficiency must be sacrificed by bleeding off some of the exhaust gas leaving the kiln to remove the volatiles by quenching. The introduction of a precalciner burning a significant proportion of the fuel makes possible the bleeding of a high fraction of the kiln gases.

3.4.4   Cooling

Once material has passed the flame, it enters a short cooling zone in which the melt solidifies with the crystallisation of C_3A, ferrite phase and dissolved silicates. Clinker leaves the kiln with a surface temperature of about 1100° and the rate of cooling between the burning zone and kiln exit has a considerable influence on its microstructure. A major objective in cooling the clinker is the rapid recovery of heat in the air subsequently used for combustion of the fuel and the resulting relatively rapid cooling prevents significant decomposition of C_3S, which is metastable below 1250°. The rate of decomposition of C_3S reaches a maximum at 1175° (Glasser, 1980) and the critical temperature interval is usually regarded as 1150–1250°, although solid solution effects may modify this somewhat. The final phases to solidify are the alkali sulfates which are formed from a melt immiscible with the major aluminosilicate melt. Alkali sulfates may also be formed on the surface of a clinker by condensation from the vapour phase.

3.5   Thermal efficiency of pyroprocessing
3.5.1   Process control

Control of the input of fuel and raw materials to a kiln is a major factor in determining its thermal efficiency. In a modern plant, continuous measurements of material flows, and of gas compositions and temperatures, are relayed to a computer which then makes adjustments mimicking a *master burner*. He relied for assessment of the success of the clinkering process on after-the-event measurements, such as clinker free lime and bulk density (*litre-weight*), the latter increasing as the consolidation accompanying clinkering proceeds, as well as the difficult measurement

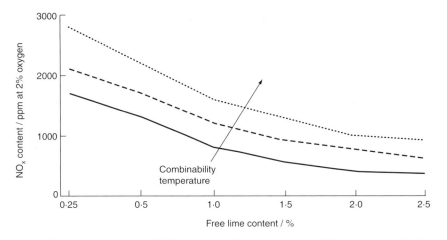

Fig. 3.10.   Relationship of NO_x content of kiln exit gas (at 2% oxygen) to clinker free lime at 3 kiln feed LSF values. Arrow indicates direction of increasing LSF (96–98–100) and combinability temperature (Lowes and Evans, 1994)

of burning zone temperature with an optical pyrometer. Measurement of the power (*kiln amps*) needed to rotate the kiln supplied him with a more immediate, although relatively insensitive, indication of the level of melt formation within it.

The development of a suitably robust chemical probe for the measurement of the NO_x (NO + NO_2) concentration in the hot, dusty gases leaving the kiln provides a sensitive method of recording changes in flame temperature and it can be linked to changes in the free lime of the clinker produced (Fig. 3.10). The nitrogen oxides are formed in small amounts as a result of molecular dissociation of O_2 and N_2 in the flame as well as from any nitrogenous material in the fuel. Measurement of NO_x, O_2 and CO concentrations at the kiln exit is used to monitor combustion conditions. In an expert system the input signals to the computer trigger output signals that seek to optimise burning conditions by changing fuel and raw feed flows. Because precise targets can not be set, predetermined small percentage step changes in these quantities are made and their effect monitored, an approach which is described as involving 'fuzzy logic'. Kaminski (1994) cited an example of typical results obtained with a kiln expert system as savings of 3% in thermal energy, 2% in electrical energy and 10% in refractory replacement costs which were coupled with a 5% increase in production. At first sight these are modest numbers, but when considered in conjunction with the scale of cement making operations, they represent valuable contributions to plant economy. To demonstrate such effects in a fluctuating system requires keeping records over prolonged periods of typical running before and after any plant modification.

3.5.2   The heat balance — process efficiency

To determine a heat balance, and therefore the thermal efficiency of a kiln-cooler system, it is necessary to measure quantities and temperatures of all materials entering and leaving the system. In addition, surface temperature measurements for the equipment are taken and standard formulae used to calculate direct heat losses by radiation and convection. The heat content of each solid and gas, referred to as its *sensible heat*, is calculated as a difference from its value at ambient temperature (usually 20°). For example, for clinker leaving a cooler at 100°, the lowest temperature likely to be achieved, the sensible heat lost is $(100 - 20)° \times 0.787 \, \text{kJ kg}^{-1}\text{degC}^{-1} = 63 \, \text{kJ kg}^{-1}$.

The calorific value of the fuel is needed to calculate the energy input. The theoretical heat required to carry out the chemical reactions can be calculated as the algebraic sum of the endo- and exothermic reactions, using the data in Table 3.2. To simplify routine computation, zur Strassen (1957) derived Equation (3.5), in which chemical formulae represent mass percentages of the oxides in the clinker produced and H_{total} is given in kilocalories per kg of clinker. Values can be converted to kJ and quoted to the nearest whole number, to well within the precision possible in a heat balance. The equation assumes an average clay mineral composition.

$$H_{\text{total}} = 2.22Al_2O_3 + 6.48MgO + 7.66CaO - 5.116SiO_2$$
$$-0.59Fe_2O_3 \tag{3.5}$$

The gases leaving the kiln or preheater, drawn through the system by a large induced-draught fan, are analysed so that the sensible heat of each component (N_2, H_2O, CO_2, CO, SO_2 and the small excess of O_2) can be calculated. The quantities involved are best appreciated from examples and those given in Table 3.4 are simplified averages of data published for the principal processes rather than optimum achievable values (Kerton and Murray, 1983).

Clearly, the absence of a knowledge of the mineral composition of the clay and aluminosilicate minerals present in a raw feed, and therefore of its true enthalpy of decomposition (H_{dec}), is a limitation on the accuracy of the result given by the zur Strassen equation. Except in sources of relatively pure kaolinite and pyrophyllite, impurities, high specific surface area, lattice defects, solid solution and interlayering of different lattice types all measurably affect the thermodynamic properties of most natural clays. Small but significant amounts of cations such as sodium, potassium, calcium and magnesium will usually be present, either adsorbed to compensate for charge imbalances created by substitutional solid solution or present in stoichiometric alkali and alkaline earth

Table 3.4.   Examples of heat balances* (after Kerton and Murray, 1983), all values are in GJ/tonne clinker

Plant type	Precalciner	Suspension preheater	Semi-dry[†]	Semi-wet[†]	Wet
Theoretical	1.78	1.77	1.72	1.76	1.75
Losses					
clinker	0.10	0.12	0.07	0.15	0.10
exhaust gases					
and dust	1.02	0.86	0.77	0.44	1.17
radiation[‡]	0.26	0.68	0.49	0.34	0.62
Evaporation of water	0.00	0.01	0.54	1.12	2.51
Total	3.16	3.44	3.59	3.81	6.15
Efficiency %	56[§]	51[§]	48[§§]	46	28

* Averages of published values.
† With grate preheater.
‡ Radiation from walls of kiln, cooler, etc.
§ Drying raw feed with kiln preheater exhaust gases increases thermal efficiency by about 5–10%.
§§ Additional drying of raw materials before nodulisation decreases thermal efficiency by 5–10%.

aluminosilicate minerals. Jøns and Hundebøl (1995) described a procedure for obtaining an averaged H_{dec} for the clay fraction of a raw feed from its content of potassium, sodium, non-carbonate calcium and magnesium (expressed as oxides) and water. They used a linear correlation of these quantities with H_{dec} that they had established for a range of well-defined clay minerals and alkali and alkaline earth aluminosilicates such as muscovite, margarite and talc.

3.5.3   Electric power consumption
A plant manufacturing cement uses a considerable amount of electric power: typical consumption ranges from around 90 kWh/t of cement for softer raw materials used in the wet process to around 130 kWh/t for the harder rocks used in a dry process. These figures include the power used in grinding the clinker to produce cement, which depends on the target specific surface (Chapter 5). In a dry process plant, raw materials and clinker grinding together account for more than about 70% of the power consumed, while burning and cooling use about 30%. The latter is used to rotate the kiln and (mainly) in driving fans. Plant design seeks to minimise resistance to gas flows, measured as a pressure drop in each section of a plant.

4. Characterisation of Portland cement clinker

4.1 Introduction

Provided that the kiln feed has been properly prepared with a regularly checked composition and fineness, as described in Chapters 2 and 3, then it is only necessary to check the adequacy of the burning process and this is done by frequent determination of the free lime in the clinker. However, as an additional check, elemental chemical composition is determined regularly by X-ray fluorescence analysis. The sulfate content of the clinker is also frequently determined so that the level of gypsum to be added in cement grinding can be calculated. In this chapter the application of a number of methods of phase characterisation will be briefly described.

In routine checks on the kiln product, the nodule size grading is watched for variation from the norm. Well nodulised clinker with a minimum of dust is needed for efficient air flow in the cooler. Nodule formation and densification also indicate a high degree of chemical combination, although if the burning temperature is too high the relationship breaks down and dust is formed. For a given burning time and temperature the clinker size range and density also depend on the silica and alumina ratios of the kiln feed. Portland cement clinker is normally black and the appearance of a paler than usual product has in the past been taken to suggest excessive chemical reduction of Fe^{3+} in the kiln. However, Scrivener and Taylor (1995) found significant reduction in some black samples of clinker and, perhaps more surprisingly, they also found that a lighter colour did not necessarily mean that reduction had occurred. Very pale clinker indicates serious underburning.

When problems arise, indicated either by the checks mentioned above or by test results obtained with cement produced from a clinker (Chapter 6), then a more fundamental investigation is necessary and clinker phase composition and microstructure are examined, usually by optical microscopy. Scrivener (1997) has provided a useful introductory account of both optical and electron microscopy and their application to cement and concrete.

4.2 Chemical analysis by selective dissolution

Free lime can be extracted from ground clinker (or cement) by treatment with a hot glycerol-ethanol mixture or by hot ethanediol (ethylene glycol). The latter is more usual and the extraction produces a solution of calcium glycollate which, as the salt of a very weak acid, can be titrated after filtration with a standard solution of HCl using a methyl red-bromocresol green indicator. Calcium hydroxide present is also extracted.

The total silicates plus free lime in a clinker can be determined by dissolving them from a ground sample using either 20% salicylic acid or 20% maleic acid in dry methanol and weighing the washed and dried residue which consists of C_3A, ferrite, MgO and sulfates (Takashima, 1958). From this assemblage it has been suggested that C_3A can be determined by dissolving it in 3% aqueous saccharose solution, although the ferrite phase is also slowly attacked and the soluble sulfates would be extracted. Selective dissolution is a considerable aid to qualitative phase analysis and characterisation by X-ray diffraction. Gutteridge (1984) used a solution of sucrose in aqueous potassium hydroxide to dissolve the aluminate and ferrite phases in order to identify the forms of alite present in cement samples.

4.3 Optical microscopy

Optical microscopy has proved very effective in characterising cement clinker because much of the microstructural detail occurs from $1\,\mu m$ upwards. Sampling of a clinker requires great care since only a very small proportion of any batch can actually be examined under the microscope and clinker from each of the range of sieved size fractions should be included. Pieces of lightly crushed clinker and fines are stabilised by vacuum impregnation with an epoxy or polyester resin which is hardened in situ.

Although a thin section of a specimen is sometimes examined in transmitted light, which makes the determination of optical properties easier, polished surfaces are more frequently examined. Nitric acid in ethanol and hydrofluoric acid vapour are common etchants. Both attack the silicates at rates dependent on their silica content, forming surface films which in reflected light produce interference colours. These provide good contrast between the di- and tri-calcium silicates, the colours depending on the thickness of the reaction layer and therefore varying somewhat with exposure to the etchant and the reactivity of the individual crystal. The use of HF also allows the (more difficult) distinction of free lime, periclase, C_3A and the sulfate phases with care. Differences in

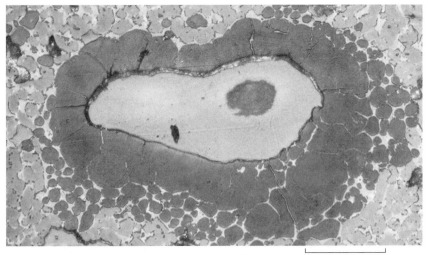

100 μm

Fig. 4.1. Photomicrograph of a polished section of Portland cement clinker. Angular alite crystals surround a cluster of belite crystals around a pore (containing stabilising resin) originally occupied by silica rich material (courtesy of Blue Circle Industries)

reflectivity and morphology are also useful aids to identification. Reflectivity is greatest for the ferrite phase.

Microscopy is used to examine clinkers qualitatively and, less frequently, quantitatively. Even without point counting of individual phases, an experienced microscopist can make a very useful assessment of the proportions of the phases present. However, the principal value of microscopy is in detecting non-equilibrium effects such as the heterogeneity of phase distribution (Fig. 4.1), porosity, crystal form and size. The size of the crystals of the interstitial phases, for example, may indicate the cooling rate experienced by the clinker (Fig. 4.2). Maki and his co-workers related the crystal size and fine structure of alite to the conditions it experienced in the burning zone of a kiln and the microstructure of belite to the cooling regime (Maki, 1994). Microscopy is also valuable in a number of other ways in the cement industry, in particular in the identification of the mineralogy of build-ups and deposits which form in kiln and preheater systems and in the examination of clinker-refractory interactions in kiln coatings.

4.3.1 Characteristics of the principal clinker phases
4.3.1.1 Alite (C_3S — density $3150\,kg/m^3$). Crystals of alite are prismatic, sometimes pseudo hexagonal, frequently having clearly defined faces. However, less regular shapes with rounded corners and faces with

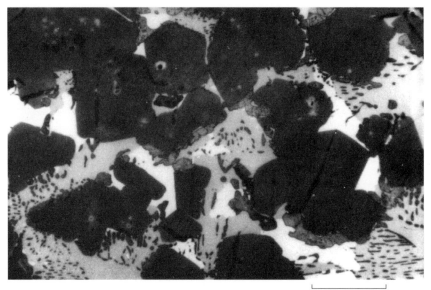

50 μm

Fig. 4.2.   Photomicrograph of a slowly cooled clinker in which ferrite (white) and tricalcium aluminate phases in the interstitial material are resolved. Small belite crystals have been precipitated during cooling (courtesy of Blue Circle Industries)

re-entrant angles are found. Internal structure observed includes lamellae and other manifestations of twinning and inclusions of belite are common. Thin sections can be used to distinguish trigonal, monoclinic and triclinic forms but to distinguish between variants with the same symmetry may require X-ray diffraction.

Sizes of crystal sections within a specimen usually range up to $100 \, \mu$m with an average around $30 \, \mu$m. In general, the size distribution in a clinker is not random. Smaller alite crystals form in the lime-rich regions while larger crystals form in the silica-rich regions of a nodule by conversion of belite (Section 2.3.1). Crystal size is also influenced by burning time and temperature, melt quantity and composition, and, if present, mineralisers such as fluoride ions.

Twin planes or fault lines may be marked by material exsolved from solid solution during cooling. For example, Fe^{3+} may be formed on cooling as striae of C_2F by oxidation of Fe^{2+} present in solid solution at clinkering temperatures. If clinker cooling is too slow, dissolved iron can significantly accelerate the decomposition of alite below 1250°. Scrivener and Taylor (1995) described the complex sequence of reactions which can occur between 1200° and 1100°. When severely reducing conditions exist in the burning zone (ca. 1400°) alite absorbs Fe^{2+} and S^{2-} from the melt, the latter being derived from the sulfur present. In the absence of oxidation during cooling, some of the alite decomposes with the formation of belite, free lime and CaO–FeO solid solution. Subsequently, an aluminium-rich ferrite and metallic iron are formed. Reduction can result in a serious fall in cement quality but is minimised by proper flame control. This problem is usually suspected if yellow-brown clinker nodules, or nodule cores, are produced and free lime rises significantly.

4.3.1.2 Belite (C_2S — density 3280 kg/m^3). Belite usually occurs as rounded crystals, frequently made up of lamellae or marked by striations indicating twinning and/or exsolution from solid solution, both effects resulting from polymorphic transformations (Section 1.3.2). Belite crystals may occur in clusters (Fig. 4.1) which, if large, indicate an unacceptable level of heterogeneity in the kiln raw feed. If a belite cluster is centred on a pore, a site formerly occupied by a silica particle is indicated. In a clinker from a coal-fired kiln, inefficient dispersion and incorporation of the ash can be detected as a localised concentration of belite, occasionally in the outer part of a nodule.

  In a slowly cooled clinker, small belite crystals separate from the melt (Fig. 4.2). When the A/F ratio is high they are also formed by interaction between alite and the melt, typically fringing alite crystals. Together with the crystal size of the aluminate and ferrite phases, these features of the clinker microstructure are indicators of cooling rate.

4.3.1.3 Interstitial phases. Tricalcium aluminate (density 3030 kg/m^3) and the ferrite phase crystallise from the melt with a degree of separation which increases with decreasing rate of cooling. The crucial cooling is that occurring in the kiln itself, that is above about 1300°. If clinker is air-quenched from the burning temperature then these individual phases may not be resolved by optical microscopy. Normally C_3A is cubic but, in the presence of alkali not combined with sulfate, alkali-containing solid solutions may form. The alkali is retained by the C_3A during cooling and part or all of the resulting solid solution will be orthorhombic (Section 1.3.3). This phase grows as elongated, lath-like crystals and exhibits birefringence.

  The ferrite phase (density in the region of 3700–3900 kg/m^3 variable as a result of variation in its composition) is distinguished by its colour in thin section and its high reflectivity in polished section. The black colour of clinker is believed to result from elements such as magnesium, titanium and silicon in solid solution in the ferrite phase and oxidation as it cools in the kiln-cooler system (Section 1.3.4).

4.3.1.4 Minor phases.   Magnesia — When the magnesia content of a clinker is greater than that which can be taken into solid solution (ca. 1.5%) periclase crystals may be observed, sometimes well formed and in clusters. More often it is observed as relicts of a dolomitic material in the kiln feed or present in the coal. *Alkali sulfates and double sulfates; calcium langbeinite* ($K_2SO_4.2CaSO_4$) — The sulfates are the last phases to crystallise during cooling of the clinker. They may also condense directly from the vapour phase. A solid state phase inversion of potassium sulfate leads to characteristic cracking. If an aqueous etchant is used, these phases are removed leaving pores. *Free lime* — Free lime is often globular as a result of attack by the melt but where interaction is limited it exists as relicts of the original calcite particles, usually surrounded by alite crystals.

4.3.2   Quantitative determination of phase composition

Using a suitably prepared polished section, point counting with a crosswire and steps of a fixed size in a grid pattern is employed with each field examined in the optical microscope. The phase observed at each step is noted and some 2000–4000 counts are made per clinker specimen. Automatic, instrumental counting may be used as in metallurgical analysis, but a number of different etches may be needed to obtain the range of contrast required for the instrument to operate satisfactorily. Counting gives volume percentages of the phases present which are converted to mass percentages using individual phase densities. The principal sources of error in quantitative microscopy derive from the difficulty in obtaining a representative sample and incorrect phase identification, especially where the crystals of the interstitial phases are very small.

4.4   X-ray diffraction

The X-ray diffraction (XRD) pattern given by a Portland cement clinker is complex and an example obtained with a diffractometer is given in Fig. 4.3. Such patterns normally take a significant time to record if the radiation counter measures the relative intensity of each reflection separately. The development of position sensitive proportional counters covering a range of several degrees of arc, without significant loss of resolution, has opened up the possibility of on-line analysis by XRD since the time to record a pattern is reduced by about two orders of magnitude. However, counting statistics must be matched to the accuracy required and in quantitative analysis a scanning detector with suitably timed counting at small intervals of degrees 2θ is usually necessary.

Identification of the phases present in clinker by X-ray diffraction is a useful supplement to microscopy, especially in identifying polymorphs,

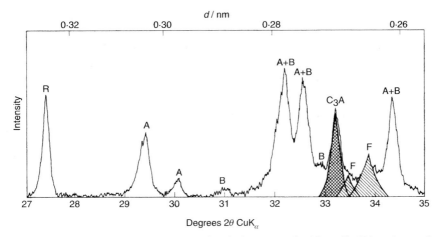

Fig. 4.3.   X-ray diffraction pattern of a clinker ground with rutile (R) as internal standard: A — alite; B — belite; F — ferrite phase (idealised peak areas are cross-hatched to indicate overlap)

although this may be difficult if several are present. It can be seen in Fig. 4.3 that the principal reflections of the clinker phases overlap considerably. This means that belite, for example, can only be identified by the weak reflection at 31.0° 2θ. All 2θ values quoted here are for weighted mean CuK$_\alpha$ radiation. Alite is identified by its reflection at 29.45° 2θ but polymorphs are best distinguished using reflections in the region of 52° 2θ.

The calcium aluminoferrite in clinker has, in the past, been considered to possess unit cell dimensions which depend on its composition within the solid solution series $C_2A_xF_{1-x}$ and a relationship between composition and the exact position of the peak at ca. 33.8° 2θ has been used to determine the value of x in the range 0.33–0.66. However, X-ray microanalysis has shown that the method is not applicable to a commercial clinker because the level of solid solution in the ferrite phase is variable, both between different clinkers and within a single clinker sample. Variations in unit cell size and cell symmetry result which are not simply related to the effect of changes in the value of x. Tricalcium aluminate is identified by its reflection at 33.25° 2θ and frequently this is the only one detectable from this phase. The single reflection from the cubic form is split if sufficient sodium is present in solid solution to reduce its symmetry to orthorhombic. The strongest reflections of minor clinker constituents may also produce small peaks.

If a cement rather than a clinker is examined, then the diffraction pattern is further complicated by weak reflections from calcium sulfate and its hydrates. Additional small peaks from calcium hydroxide and calcium carbonate formed by contact with moist air, and from calcite, quartz and clay minerals, possible impurities in the gypsum used, may be present. Heating a sample to 500° to dehydrate the hydrous compounds simplifies the pattern somewhat, but a temperature necessary to decompose calcite (ca. 900°) can not be used because of the risk of decomposing alite.

4.4.1   Quantitative X-ray diffraction analysis (QXDA)

QXDA of a mixture of phases is based on the fact that the intensity of a particular reflection (area under a peak) is proportional to the volume fraction of the phase producing it. The proportionality constant is determined by the crystal structure of the phase, the sample and equipment geometry, and the level of X-ray absorption occurring in the mixture. A weighed proportion of an *internal standard* is interground with the sample to eliminate the 'practical factors' associated with specimen preparation and exposure to x-radiation in the diffractometer. Ratios of the intensities of the selected reflection from each component of the clinker (or cement) to a chosen reflection for the internal standard are determined and calibration curves, prepared with synthetic reference samples of each of the clinker minerals, are used to convert intensity ratios for the unknowns into mass fractions. The internal standard chosen should provide a reflection close to those of the clinker minerals but not overlapping them. Rutile (TiO_2) or silicon are frequently chosen and the reflection used for rutile is indicated in Fig. 4.3.

Cooperative inter-laboratory exercises have revealed the limitations of this conventional single-peak method in QXDA of clinker and cement, although it can give acceptable results for C_3A for which the peak at 33.25° 2θ is used. Aldridge (1982) reported results obtained using chemical (Bogue), microscopic and single peak QXDA for the phase analysis of six commercial clinkers and the corresponding cements in eleven laboratories. Between-laboratory standard deviations found for alite contents, ranging from 45 to 75% in the six clinker samples, and the cements (values in brackets) were: Bogue 3.6 (2.2); microscopy 3.4; QXDA 8.9 (7.2). For C_3A contents (range 5–10%) in the same clinkers and cements these methods gave: 0.8 (0.4); 2.0; and 2.1 (2.1), respectively. Within-laboratory standard deviations were mostly from 0.4 to 0.8 times between-laboratory values. The only between-laboratory standard deviations considered acceptable were those for chemical analysis (Bogue). The higher standard deviations observed for clinkers than for cements were attributed to greater sampling errors.

The principal sources of error in conventional QXDA (other than equipment and sampling) are considered to be:

(a) reference clinker phase used in the preparation of a calibration curve does not have the same composition/structure as that in the sample being analysed and also there may be more than one polymorph present in the sample being analysed;

(b) difficulty in locating the base line in peak area measurement;

(c) the use of a weak peak for a phase such as βC_2S because its strong peaks overlap with those of other phases;

(d) poorly standardised and over-grinding which can affect clinker constituents and internal standard differently;

(e) preferred orientation of crystallites;

(f) phase segregation during sample preparation.

Attempts to overcome the first three problems have involved a change from single to multi-peak (whole pattern) methods. They require accurate qualitative identification of the phases present in the unknown. A computer is used to create and match by a least-squares procedure weighted data for the reference phases identified and the digitally recorded diffraction pattern from the sample being analysed. Refinements can be introduced to allow for line broadening effects, such as crystallite size, and for background radiation as a function of 2θ. Gutteridge (1984) created a database of diffraction patterns recorded over the range 24–39° 2θ for 10 alite, 6 dicalcium silicate, 4 C_3A and 10 ferrite compositions, together with a further 19 patterns for minor phases which can be present. Most of these have been incorporated into the JCPDS files (Joint Committee on Powder Diffraction Standards).

A second method uses published basic crystallographic data, for the phases identified in a sample of clinker or cement, to calculate the proportions of each which give the best fit of calculated and observed diffraction patterns over a wide range of 2θ values. The method of breaking down (deconvoluting) composite peaks was originally devised by Rietveld (1969) for neutron diffraction and adapted subsequently for QXDA. Its wider use has been made possible by the increasing application of computers since about 1990. The essential feature of the method is the simultaneous, interactive calculation of a least-squares fit and diffraction pattern refinement (Young, 1993). Programmes available permit the introduction of line broadening effects and the variation in background intensity. To obtain meaningful results by this technique requires high quality equipment and a high level of theoretical and experimental technique.

Taylor and Aldridge (1993) obtained root mean square deviations between Rietveld and optical microscopy in the quantitative analysis of six cements as: 2.1% for alite, 3.1% for belite, 1.7% for the aluminate, and 1.3% for ferrite. Neubauer et al. (1997) reported acceptable correlations between the contents of alite, belite, free lime and C_3A + ferrite determined by Rietveld analysis, and those obtained by microscopy for twelve samples of commercial clinker taken from a single production line. However, poorer correlations were obtained for five samples from a second plant using waste oil and sewage as partial coal replacements. Since computers are capable of making the calculation within two or three minutes, the potential of the method is being examined for quality and possibly process control in cement plants (Möller, 1998).

4.5 Electron microscopy

Transmission electron microscopy (TEM) is regarded as a research method of value in the investigation of composition and defect structure of clinker particles at the highest levels of resolution and electron diffraction can be used to confirm phase identity. In the examination of microstructure, extremely thin specimens are required and they are produced by ion beam thinning of sections prepared in the usual way (Groves, 1981). TEM has been used to examine defects in alite crystals (Hudson and Groves, 1982) and twinning in belite (Groves, 1982). It has also been of considerable value in the examination of hardened cement pastes (Section 8.2) although the sample must be liquid nitrogen cooled at the ion beam milling stage and extreme care is required in minimising exposure to the electron beam in the microscope to avoid damage (Richardson and Groves, 1994). Scrivener (1997) pointed out that it is vital that samples examined by TEM are characterised by other methods to ensure that the small amount of material which can be examined is representative.

The initial application of scanning electron microscopy (SEM) in the 1960s and 1970s to cement-containing materials made use of the secondary electrons emitted from the surface 10 nm of a sample when it is exposed to an electron beam. In the most modern instruments, resolution is of the order of nm compared with angstroms in TEM. A particularly wide range of magnifications (from 20 up to 200 000) is available, making possible the examination of the topography of fracture surfaces of hardened cement pastes, mortars and concretes at several levels (Fig. 8.1). Where the extreme drying conditions of very high vacuum are undesirable, an environmental cell or an environmental microscope (ESEM) can be used. An example described by Donald (1998) employs the differential pumping principle. While the electron gun is maintained at 10^{-7} torr, the other sections of the microscope are maintained at progressively higher pressures with the specimen held at around 10–20 torr.

4.5.1   Backscattered electron (BSE) imaging

The development in the 1980s of detectors for the more energetic backscattered electrons in the scanning electron microscope made it possible to obtain images of the microstructure of clinkers and partially hydrated systems which quantitatively reflect phase content. Sections to be examined must be polished to provide uniformly flat surfaces after impregnation and fixing of weak material and particulates in a resin. Samples are coated with carbon to prevent charge build up. BSE images are not unlike optical ones, although no etching is required and no colour is involved. The intensity of backscattering by a phase is determined primarily by its physical density and approximately by the mean atomic number of the elements it contains.

Phases are differentiated by the level of greyness in their BSE image, ferrite being the brightest with cracks and pores, at the other end of the scale, black (Fig. 4.4). Contrast between C_3A and the ferrite phase is such that they can be distinguished in the interstitial material but the former is not readily distinguished from belite. However, grain boundaries are usually visible as a result of a small amount of phase contrast. Hydration products can be observed as darker than the anhydrous grains they surround. Automated image analysis can discriminate most of the phases and pores in a BSE image and in hydrated material can be used to determine degree of hydration or the pore size distribution in a hardened paste (Section 8.2.3).

4.5.2   X-ray microanalysis

Compositional variations in a specimen can be examined by X-ray microanalysis. This technique is based on the measurement of the intensities of characteristic elemental lines in the X-ray spectra excited by the electron beam, using one or more wavelength or energy dispersive detectors. Quantitative analysis requires a flat specimen surface and calibration using minerals of known composition as standards. To minimise errors introduced by X-ray fluorescence and absorption effects, standards must have a composition as similar to the unknown as possible but synthetic C_2S and C_3S would be too reactive as even superficial hydration and carbonation would seriously affect results. Synthetic gehlenite (C_2AS) is a possible compromise composition.

The principal limitation of the technique is the electron probe size and penetration, that is the probability that with very small crystals (< around $5 \mu m$) more than one phase may be excited. In such cases plots of one atomic ratio against another can prove valuable aids in interpretation (Fig. 8.2). X-ray microanalysis is also an aid to interpretation of a microstructure. Scanning a sample to record changes in the intensity of

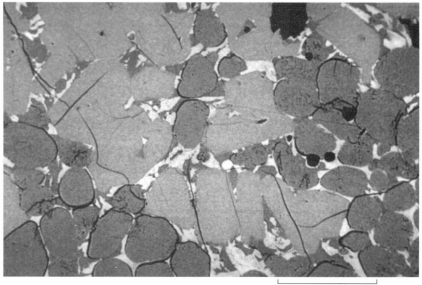

50 μm

Fig. 4.4.   Backscattered electron image of a well burned clinker. Crystals of belite are distinguished from C₃A by their characteristic rounded form (courtesy of Blue Circle Industries)

the X-rays emitted by an individual element produces a 'dot map' of the microstructure, in which the presence and concentration of an element is indicated by the density of dots. Such maps indicate the distribution of elements in solid solution and this may reveal grain boundaries for which contrast in reflectivity is inadequate.

The electron probe microanalyser is essentially an SEM primarily dedicated to X-ray microanalysis, while an SEM is equipped with an energy dispersive X-ray facility but is designed to optimise imaging. A transmission electron microscope can also be used for X-ray microanalysis and usually this is referred to as analytical electron microscopy. It has the highest resolution (a volume ca. 200 nm across by ca. 200 nm thick) but specimen preparation is much more time consuming. In addition, absolute atom contents, as opposed to atom ratios, are difficult to determine in the TEM.

4.6   Concluding remarks

A major objective of clinker characterisation is to support a manufacturing unit when normal quality control reveals a problem. For example, the problem might be an unexpected increase in combinability temperature, revealed as a significant increase in clinker free lime and prompting the use of more fuel than normal. If this occurrence was not accompanied by any obvious change in raw meal composition and fineness, the phase composition and microstructure of the clinker would be examined, as well as the detailed chemistry and texture of the raw materials. Often, as quarrying proceeds, there are gradual changes in the composition of the material extracted and records are kept to identify trends. When the composition and fineness of a raw mix for a new works is being optimised, the detailed characterisation of experimental clinkers is a valuable aid.

Many early studies of the relationship between raw materials, the processing conditions employed, and the characteristics of the clinker and quality of the cement produced, have not been of general applicability because the chosen variables and the range examined have only been relevant to a particular manufacturing unit. Fundamental factors influencing hydraulicity, such as polymorphism, solid solution and crystal defect structure, including that introduced by grinding, have not usually been taken into account. The potential number of variables is considerable, but as Glasser (1998) has pointed out, there has been a growing interaction between fundamental studies and empirical investigations seeking to optimise product quality with a given set of raw materials.

5. Grinding and fineness of cement

5.1 Cement milling

After it leaves the cooler, clinker is conveyed to a covered store in which some blending may be possible. Cement is produced by grinding clinker and gypsum, usually in a tube mill. This is divided into two or three chambers by means of slotted partition walls (diaphragms) which permit the forward movement of cement but retain the size-graded grinding media (Fig. 5.1). Milling is continuous and the residence time of the material in the mill, and therefore the fineness of the cement, depend on the rate at which clinker and gypsum are introduced. A large mill drawing 4500 kW, 4.6 m in diameter and 14 m long, would contain about 280 t of steel balls with diameters from 90 mm in the first chamber down to 15 mm in the last chamber. A mill may operate either on open circuit, that is with the product going direct to a storage silo, or on closed circuit with the product being conveyed by air or a mechanical elevator to a separator (classifier) from which coarse material is returned for further grinding.

The wall of a ball mill lifts the media as it rotates and at a certain height they fall to grind the cement; the mill must not rotate above the critical speed producing a centrifuging action. The shell of the mill is protected by liner plates which may have a rippled profile to optimise lifting, since slippage of the media results in energy loss. Efficiency is rated in terms of the surface area produced per unit of electrical energy consumed and a 'standard energy requirement' of $1.15 \times 10^4 \, \text{m}^2/\text{kWh}$ may be used as a basis of comparison. Energy consumption is approximately linear up to about $300 \, \text{m}^2/\text{kg}$, above which it increases progressively per unit increase in surface area as cushioning becomes more serious. In everyday running, the residues in the mill product on $90 \, \mu\text{m}$ and $45 \, \mu\text{m}$ sieves are used to monitor mill performance, which may decline as a result of media wear, for example. Increase in such residues at a given surface area will result in a change in the compressive strength/curing time relationship of the cement. A comparison of the effects of open and closed circuit milling on these residues can be seen in Fig. 5.2.

Fig. 5.1.   Interior of a ball mill showing liner plates and a diaphragm (courtesy of Christian Pfeiffer)

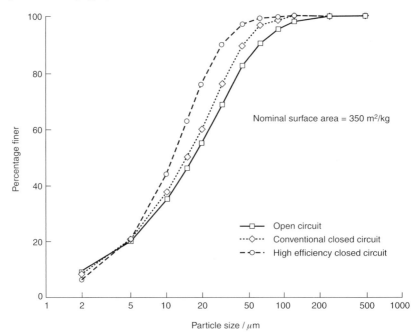

Fig. 5.2.   Effect of classifier efficiency on the particle size grading of ordinary (42.5N) Portland cement (Moir, 1997)

Power consumption in ball milling Portland cement is of the order of 45 kWh/t for a surface area of 360 m^2/kg. This may be reduced by employing a closed circuit system, the saving of 2–5 kWh/t depending on the efficiency (fan power requirement) of the separator. The principal variables to be considered in optimising energy consumption in a ball mill include: the speed of rotation of the mill, its ball size-grading and loading, and the design of its lining. The use of a grinding aid reduces energy consumption, especially with higher surface area products. A mill is designed on the basis of the throughput required, using data for the grindability of the clinker determined in the laboratory (Section 2.4).

The ideal way to grind a material would be to break each crystal or aggregate of crystals separately by simple cleavage. The energy consumed is then the surface energy created plus that lost to the fragments and media as heat. However, those milling systems which most efficiently keep the particles separated for grinding, such as the roller (vertical spindle) mill or the roll press (Figs 3.4 and 5.3), are the most susceptible to wear with a hard, abrasive material like clinker. This resulted in serious maintenance problems in early versions of these mills

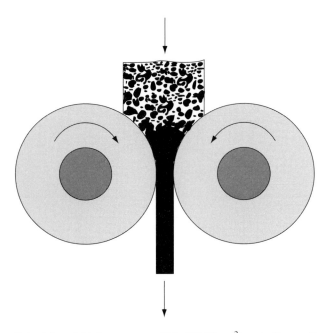

Fig. 5.3.   Principle of high pressure (50–400 N/mm^2) grinding roll; the cake produced is crushed and the fines separated (Moir, 1997)

before replaceable, wear-resistant alloy surfaces were developed. Consequently, although a ball mill wastes energy in multiple impacts (cushioning), it has remained predominant in cement grinding. Capital costs usually preclude the complete replacement of a grinding plant in an established works.

Initial grinding of large clinker nodules in a ball mill is particularly inefficient. In existing plants the introduction of a roll press for preliminary or semi-finish grinding, with finish grinding in a ball mill, has proved a cost effective way of significantly increasing both energy utilisation and throughput, the latter making it possible to maximise use of low tariff (off-peak) electricity. Improvements in both are even greater in raw materials grinding (Ellerbrock and Mathiak, 1994). This combination of grinding techniques has the advantage of avoiding two adverse effects when a roller mill or roll press is used for finish grinding, namely an increase in the *water demand* of the cement and the possibility of an unacceptable reduction in initial setting time. The former is ascribed to the narrowing of the particle size distribution resulting in an increased voidage in the cement (decreased bulk density) and the latter to the production of coarser gypsum particles and a lower degree of dehydration resulting from a lower grinding temperature (Section 7.3).

The relatively recently introduced Horomill (horizontal roller mill), which is suitable for the finish grinding of cement clinker and raw materials, is essentially a tube mill in which a cylindrical roller constitutes the grinding component (Fig. 5.4). Cordonnier (1994) describes the performance of the first industrial (25 t/h) installation of this mill in Italy. For a similar capacity to a ball mill, it has a slightly smaller diameter and

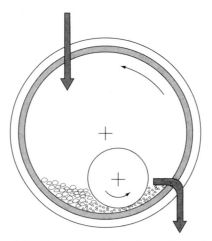

Fig. 5.4.   Principle of the Horomill — directions of rotation and material flow are indicated (Ellerbrock and Mathiak, 1994)

is only one-third of the length. Energy saving was quoted as 30–50% with wearing surfaces having a satisfactory life. For a cement with a surface area of $360\,m^2/kg$, an energy consumption below $30\,kWh/t$ can be expected. Cement produced by the Horomill had a similar particle size distribution and similar physical properties to one produced in a ball mill.

5.1.1   Factors influencing the grindability of clinker

The grindability of clinker depends on its chemistry and on the conditions it experiences in burning and cooling. Hard burning and high melt content resulting from a low silica ratio increase initial grindability since they result in a clinker with a low porosity. (Grindability increases with increasing difficulty of grinding — Section 2.4.) Maki *et al.* (1993) observed that grinding was impaired in clinker containing clusters of belite crystals. After most of the larger aggregates of crystals have been broken, the fracture properties of the individual phases assume greater importance, although it must be remembered that a majority of the final cement is made up of multiphase particles. Hardness of a crystal is less important than its brittleness in comminution and since alite cracks much more readily than belite in a microhardness measurement, clinkers with a high lime saturation (and substantially complete chemical combination) can be ground more readily than those with a low lime saturation.

Hornain and Regourd (1980) found that the grindability of a clinker sample sintered to a density of $3000\,kg/m^3$ was determined by its fracture energy and the size of the microcracks present. The number and size of the latter could be related to the cooling regime experienced by the clinker. They measured fracture energies in the range $12–20\,J/m^2$, using notched, sintered prisms of clinker. From measurements of the impression made in each clinker phase by a Vickers micro-indenter and the size of the cracks radiating from the indentation, they calculated values for a brittleness index: C_3S 4.7; C_3A 2.9; C_2S and C_4AF 2.0.

Scrivener (1989) examined cement particles by BSE imaging and X-ray microanalysis and found that fracture of alite crystals predominated in polymineralic clinker particles rather than fracture along phase boundaries. Many smaller particles had surfaces rich in interstitial phases. Bonen and Diamond (1991) used the same techniques to examine two chemically very similar cements, one produced in a ball mill the other in a roller mill. The latter contained more nearly isodimensional particles and a significantly smaller proportion of the finest particles present in the sample from the ball mill. The particles in the ball-milled sample also exhibited much greater surface roughness. X-ray microanalysis revealed that the surfaces of the particles in the two cements differed in composition, that from the roller mill having a higher content of belite and interstitial material, apparently reflected in a significant decrease in heat release in the first day of hydration.

5.1.2   Minor additional constituents

The current British Standard (BS 12: 1996) follows the European pre-standard (Section 6.1 in permitting up to 5% of a *minor additional constituent (mac)* to be ground with clinker as part of the *nucleus* of a Portland cement, that is excluding the gypsum. Unlike the European specification, it limits the materials which can be used to one or more of a natural pozzolana, blastfurnace slag, or pulverised fuel ash but not if a blended cement is being manufactured with one of these as a main constituent (Section 9.3) or if a *filler* such as ground limestone is being added. The term 'filler' covers any inorganic natural or artificial material which, owing to its particle size grading when ground, enhances the physical properties of cement without any detrimental effect on concrete durability. Moir (1994) concluded that ground limestone is likely to be a preferred mac for practical and economic reasons. It aids the control of cement workability and strength development and inhibits bleeding.

5.1.3   Addition of gypsum

Gypsum is added as a set regulator in an amount required to raise the previously determined SO_3 content of the clinker to the target value for the cement. Gypsum is more easily ground than clinker so that even a modest fall below target significantly reduces mill output, by increasing the time necessary to reach the desired surface area. The required addition is usually taken as that which produces the highest 28-day concrete strength determined by standard procedures (Section 6.4). The experimentally found optimum SO_3 is dependent on the chemistry of the clinker so that the level in commercial cements varies with source, an upper limit being set to protect consumers from excessive addition. Too high a level introduces a risk of concrete volume instability caused by the formation of the sulfoaluminate, ettringite, after the concrete has hardened.

Grinding aids such as diols (glycols) or triethanolamine at levels of about 0.1% have been found to reduce the energy needed in milling to reach a given fineness and their use is now widely permitted. They are considered to improve the flow of cement in the mill and the separator and to function by adsorption on to freshly formed surfaces as cleavage occurs, thus inhibiting re-agglomeration. Gypsum itself is sometimes referred to as a grinding aid, possibly only because it is softer than clinker. However, the water vapour produced as gypsum is heated by the action of the grinding media would certainly be strongly adsorbed on cleavage surfaces.

The extent to which calcium sulfate dihydrate is dehydrated in a cement mill depends on the temperature and relative humidity existing in it. The *hemihydrate* formed does not have the precise composition implied by its name because compositions covering the range $CaSO_4.0.01$–$0.63 H_2O$ are obtained. The name *soluble anhydrite* is sometimes used for the low end of the range. These compositions can be interconverted simply by a change in relative humidity or temperature since water molecules can enter or leave the porous structure. In contrast, conversion to the dihydrate involves its dissolution in water from which the dihydrate crystallises as the stable phase at temperatures below about 43°.

In order to minimise the risk of producing a false setting cement, milling should not be carried out at too high a temperature (test methods Section 6.5; chemistry Section 7.3). Mill temperature is limited by supplying the mill with cool clinker, by cooling the recirculating air and in some cases by using an internal water spray. This must do no more than cool evaporatively, of course, and the flow of water is controlled by monitoring the temperature of the cement leaving the mill and the loss on ignition of the final product. Even a modest rise in the moisture content of a cement can cause problems for the cement maker, because it raises the relative humidity in the silo in which it is stored. The increased humidity enhances adhesion between cement particles and lumps may form. Alsted Nielsen (1973) pointed out that this can happen if too much undehydrated gypsum is present in a cement and it enters a silo at temperatures above about 70°.

5.2 Fineness of cement

The fineness of a cement is a major factor influencing its rate of hydration, since the chemical and physical processes involved occur at its interface with water. Ordinary (BS 12 strength class 42.5N) Portland cement is usually ground to a surface area of 330–380 m^2/kg and rapid hardening (strength class 52.5N) to 400–450 m^2/kg. Values are thus principally determined by the strength class being produced (Table 6.1). The principal methods of measuring surface area will only be briefly described here, since detailed descriptions of procedures and equipment are given in the standards of a number of countries. The value obtained is sensitive to the method employed, which should be quoted when reporting a result. The Lea and Nurse reference method and the Blaine method are described in BS EN 196 part 6: 1989. The latter is used routinely in the UK. Another way of expressing the fineness of a cement is by its particle size distribution (grading) and an indication of methods employed and its relevance to strength development is included in this section.

5.2.1   Determination of surface area

The weight-specific surface area of cement (S_w) is usually determined by an air permeability method, the principles of which are discussed by Allen (1995). The relationship between S_w and the measured resistance to flow of a powder bed for a specific surface in the range in which laminar (viscous) flow occurs (250–500 m²/kg) is given by the Carman–Kozeny equation (5.1):

$$S_w^2 = \frac{N}{\rho(1 - \epsilon)} \frac{\epsilon^3 A \Delta p}{\eta Q L} \tag{5.1}$$

where A, L and ϵ are the bed cross-sectional area, thickness and porosity; η is the Stokes viscosity of air and Q its rate of flow; ρ is the density of the powder and Δp the drop in pressure across the bed; N is a dimensionless constant dependent on the units chosen.

In the Lea and Nurse apparatus, dry air is passed continuously at constant pressure, first through a compacted cylindrical bed of cement (25 mm diameter, 10 mm deep) and then through a length of capillary tubing. A manometer is used to measure the drop in pressure across the bed of cement (h_1) while a second manometer measures that across the capillary which acts as a flowmeter. For the apparatus normally employed for cement, Equation (5.1) becomes:

$$S_w = \frac{k}{\rho} \sqrt{\left(\frac{h_1}{h_2}\right)} \tag{5.2}$$

where h_1/h_2 is the average of the ratios of the two manometer readings at two air flow rates and k is a constant dependent on the dimensions of the apparatus.

The density of cement is determined with a pyknometer, using kerosene and a vacuum pump to ensure complete removal of air. Densities of Portland cements usually lie in the range 3050–3250 kg/m³. The mass of cement used in the apparatus is adjusted to produce a bed with a porosity of 0.50. Details of the apparatus, procedures and physical data, such as the viscosity of air at different temperatures which is needed in the calibration of the flowmeter, are given in the UK National (informative) Annex B in BS EN 196-6.

The European Standard permeability method of measuring the specific surface of cement is based on a principle developed independently in 1943 by Rigden and Blaine. The constant volume methods they introduced make use of the fact that, with equipment of suitable dimensions, the time t required to pass a fixed volume of air through a bed of cement of standard porosity is related to its specific surface by the simple relationship (5.3). In these methods the pressure decreases as air passes through the specimen.

$$S_w = K\sqrt{t} \tag{5.3}$$

where K is a constant for the apparatus determined using a reference Portland cement of known specific surface (available from the National Institute for Standards and Technology, Washington DC, USA).

The simplicity of constant volume permeameters makes them suited to general use in the cement industry. The specific surface of the reference cement used in their calibration must have been determined recently by the Lea and Nurse absolute method because small but measurable changes occur when cement is stored, especially when it is freshly ground. Exposure to moist air must be minimised because this produces larger, more rapid changes.

In one version of the constant-volume permeameter, kerosene in one limb of a U-tube is raised to produce a head of about 100 mm. This is then allowed to fall, forcing air through the cement which is compacted in a cell attached to the second limb. The time is noted for the meniscus to fall between two marks on the U-tube which define the fixed volume. Automated versions exist in which surface area is directly indicated.

Another method for determining the specific surface of cement, following Wagner (1933) in using light extinction, is described in the American standard, ASTM C115. The cement is dispersed in kerosene. A column of the suspension is allowed to settle and the intensity of light transmitted is measured as a function of time and depth below the surface of the column. In this standard the results are used to determine specific surface area although they could provide a particle size distribution. Because the Wagner procedure assumes that all particles with a diameter less than 7.5 μm have an average diameter of 3.8 μm, it gives a lower specific surface than an air-permeability method; it underestimates the considerable contribution which the finest particles make to total specific surface.

A cement with a 'Wagner' surface of 180 m^2/kg might have a 'permeability' surface of 320 m^2/kg. If the specific surface area were determined from nitrogen-adsorption measurements at 77K, using the BET calculation procedure (BS 4359, Part 1), then a value in the range 800–1000 m^2/kg would probably be obtained. The main reason for the much higher value is that a significant proportion of the total surface of cement lies in pores accessible from one end only, for example in the porous hemihydrate formed from gypsum in the mill. The surface area in 'blind pores' is not determined in the permeability methods described here because they do not contribute to flow, but the nitrogen adsorption surface area includes them.

5.2.2   Particle size distribution

Until recently a complete particle size distribution was only determined for research purposes but the availability of relatively low cost and reliable laser granulometers has made possible their use in quality control. The techniques which have been employed to determine the particle size distribution of cements are those generally applicable to powders: sedimentation in a non-aqueous liquid such as ethanol, using an Andreasen pipette or a sedimentation balance; the Coulter Counter; X-ray or light extinction and light scattering methods. Principles of the methods are discussed by Allen (1996).

Particles larger than about 45 μm are usually determined on standard sieves and the finer particles by one of the above techniques. The results shown in Fig. 5.2 were obtained in this way, a sedimentation X-ray extinction method being used for the finer particles. Ethanol, with a small amount of calcium chloride dissolved in it, is a suitable suspending medium for sedimentation and light scattering measurements. Laser granulometers, which allow the measurement of the very low angle, particle size dependent, scattering of light by a suspension have been increasingly used in the cement industry since the late 1970s. A critical evaluation of a number of commercial instruments was made by Geurts *et al.* (1992).

Numerous attempts have been made to correlate the compressive strength development produced by a cement with its particle size distribution. A generally held view has been that the 3–30 μm fraction makes a major contribution to 28-day strength. Kuhlmann *et al.* (1985) obtained results with different size fractions which supported this and noted the importance of the <3 μm range in achieving a high 1-day strength.

A particle size distribution is usually recorded as a cumulative curve (Fig. 5.2) and for most Portland cements (and many other ground substances) this can be represented by an equation proposed by Rosin and Rammler:

$$R(d) = 100 \exp(-d/d')^n \qquad (5.4)$$

or in linear form:

$$\ln\ln R(d)/100 = n(\ln d - \ln d') \qquad (5.5)$$

where $R(d)$ is the percentage of particles (by mass) with an equivalent spherical diameter greater than d, and d' the *position parameter*, the diameter at which the residue is 36.8%. The percentage coarser than each chosen value of d is then plotted against particle size on a Rosin-Rammler-Sperling-Bennett (RRSB) diagram, an example of which is given in Fig. 5.5. From this linear plot the size distribution of a cement

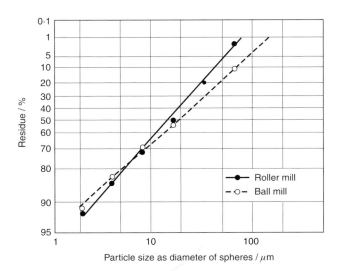

Fig 5.5. Effect of mill type on the particle size distribution of cement produced, plotted on an RRSB grid. A laser granulometer was used to determine the size distribution of two cements with a Blaine surface area of 280 m²/kg. Slopes of the plots were 1.11 and 0.90 (Osbaeck and Johansen, 1989)

can be characterised by two parameters: n the slope, which is a measure of its dispersion, and d' which is an indication of its fineness. Fig. 5.5 contains plots for two cements ground in different mills. Plots for cements ground in the same mill for different times have the same slope but are displaced with respect to the particle size axis.

Using these parameters, Kuhlmann *et al.* examined the influence of particle size distribution of laboratory prepared cements on the growth in strength of standard mortar specimens. For cements with a constant specific surface area, they found an increase in strength with increasing slope of the RRSB plot (Fig. 5.6), while with cements having an equal slope, strengths decreased with an increase in position parameter, that is with increasing coarseness as would be expected.

The particle size distribution of a cement also influences the amount of water needed to produce a paste with an acceptable standard consistency (Section 6.3). The *water demand* of a cement is determined by the efficiency with which the particles are packed in it, since water must first fill the voids in the powder before it contributes to flow. In addition, some of the added water will react with the cement. Thus grinding will affect water demand through the effect it has on particle size distribution, particle shape and roughness, as well as the initial reactivity of the phases exposed as fracture takes place.

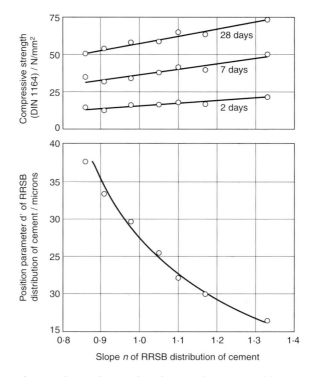

Fig. 5.6.   Correlation of particle size distribution characterised by parameters of RRSB plots with standard mortar strengths (DIN 1164) for cements with a Blaine surface area of 220 m²/kg (Kuhlmann et al., 1985)

6.   Tests of cement quality

6.1   Introduction

The assessment of cement quality relies primarily on direct performance tests because of the complexity of the factors influencing its rate of hydration and its hydraulicity. It was seen in the last chapter that the value obtained for the specific surface area of cement is particularly method-dependent, so that a prescribed procedure must be followed and named when referring to the result obtained. A similar constraint applies to the determination of the hydraulicity of cement by incorporating it in a concrete or mortar.

Committees of specialists, representing all interested parties (manufacturers, consumers, government and academic institutions), have in some countries produced national specifications and test methods for the assurance of cement quality. In other countries, British Standards or those published by the American Society for Testing and Materials are used. Standard test procedures have also been published by the International Standards Organisation and by the Comité Européen de Normalisation (CEN). Those developed by the latter, Methods of Testing Cement: EN 196, have been published with useful National Annexes by the British Standards Institution as parts of BS EN 196. CEN is currently finalising a specification for cements commonly available in Europe. A pre-standard is voluntary (ENV 197) but drafts have formed the basis of the present British Standard, BS 12: 1996. It should be noted that standards and specified test procedures are regularly reviewed and revised when deemed necessary. BS 12 was first published in 1904. ASTM standards are published annually in book form; cement and concrete specifications and testing are included in Volume 04.01.

A standard specification lays down the chemical, physical and performance characteristics required of a cement for it to be sold as conforming to the standard. In the current European approach, a programme of product sampling (defining minimum frequency and method) is indicated for the manufacturer who must sample at the point the product leaves a works and

Table 6.1.   Requirements of BS 12: 1996

Property	Strength class	Requirement: % m/m cement*	Deviation limit[†]
Loss on ignition	All	≤ 3	+0.1
Insoluble residue	All	≤ 1.5	+0.1
Sulfate as SO_3	All	≤ 3.5	+0.1
Chloride	All	≤ 0.10	+0.01
Initial set	32.5N, R 42.5N, R	≥ 60 min	-15 min
	52.5N, 62.5N	≥ 45 min	
Soundness	All	≤ 10 mm	+1 mm

Compressive strength/Nmm^{-2}

Class	Early strength		Standard strength		Deviation limit[†]	
	2 days	7 days	28 days		All classes	
			Lower	Upper	Lower	Upper
32.5N	–	≥ 16	≥ 32.5	≤ 52.5	2(7)d 28d	Not specified
32.5R	≥ 10	–			-2.0 -2.5	
42.5N	≥ 10	–	≥ 42.5	≤ 62.5		
42.5R	≥ 20	–				
52.5N	≥ 20	–	≥ 52.5	≤ 72.5		
62.5N	≥ 20	–	≥ 62.5	–		

*   Cement defined as clinker + gypsum + grinding aid (if any).
†   Maximum deviation permitted for individual results.
R: high early strength sub-class.

employ the test methods laid down in EN 196. Two statistical procedures, and values for the necessary statistical parameters, are provided for the assessment of test results obtained against the specification (*manufacturer's autocontrol*). Criteria for conformity take the form of specified *characteristic values* for properties which can only be breached by a defined number of test results in a given set (test period). They are derived from probability theory using a defined, low risk of acceptance of a batch not meeting the required characteristic value. These procedures are described in detail by Brookbanks (1994) with helpful worked examples. In addition, for some properties, *limit values* are also specified and no individual autocontrol test result must fall outside these.

The British Standard BS 12: 1996 specifies chemical, physical and performance requirements for Portland cement (Table 6.1) following ENV 197. In addition to characteristic values, the British Standard specifies *acceptance limit values* for certain properties which are somewhat more stringent than the limit values in ENV 197. They can be used with results obtained for single samples by a customer or independent test laboratory. Maximum permitted deviations above or below the stated acceptance limits for individual results are also specified (Table 6.1).

6.2 Chemical composition

The compositional requirements specified for Portland cement cover both clinker and cement. The test methods to be employed are those described in BS EN 196. A detailed discussion of these is given by Taylor (1994). The compositional requirements for clinker are: $C_3S + C_2S > 66.7\%$; C/S > 2.0; MgO $< 5.0\%$. They are comfortably met in the UK. The requirements and acceptance limit values specified for cement in BS 12: 1996 are given in Table 6.1. The limit for chloride ion content is necessary to reduce the risk of corrosion of steel in reinforced and prestressed concrete. Limits for loss on ignition and insoluble residue protect the consumer from a product which has suffered either excessive exposure to the atmosphere during storage or contamination.

6.3 Setting times

These are the times after completion of mixing at which a neat cement paste presents specified resistances to the penetration of a needle. The principle variables influencing penetration are the water content of the paste, the temperature, the load on and dimension of the needle and, of course, the reactivity of the cement. The needle employed (diameter 1.13 mm, total load 300 g) is named after Vicat (1828). It is released at the surface of the hydrating paste at intervals until it penetrates only to a point 4 ± 1 mm from the bottom of the standard mould (Fig. 6.1). When the paste has attained this degree of stiffness it is said to have reached *initial set*, for which a minimum value is specified in BS 12 (Table 6.1). A second similar needle with a concentric ring attached can then be used to determine *final setting* time, although a maximum value for this is no longer specified in the British Standard. Final set is reached when the needle makes an impression on the surface of the paste but does not penetrate the 0.5 mm necessary for the ring to mark the surface.

The higher the water content of the paste, the longer it will take for the cement hydration products to form a structure with the chosen resistance to penetration. BS EN 196 does not, however, specify a fixed water/cement ratio. Instead, pastes are examined at a range of ratios to establish that needed to produce a paste into which a 10 mm dia. plunger, which is held in the apparatus used for the Vicat needle, penetrates to 6 ± 1 mm from the bottom of the same mould. This paste is described as having *standard consistence* and since the result is sensitive to shear history, the mixing procedure is specified.

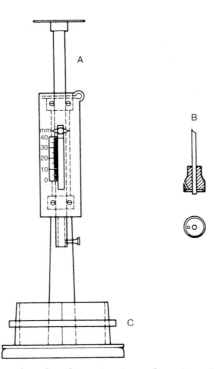

Fig. 6.1. Apparatus for the determination of setting times and standard consistence (BS 4550): A—holder for Vicat needles and consistence plunger, B—final set needle, C—standard mould

6.4   Compressive strength

The most important test of cement quality involves the determination of the compressive strength it produces in a mortar or concrete. In the past, a specified concrete mix was usually tested in the UK using British Standard 4550 although this also gave a procedure for a mortar. The USA and many European countries favoured mortar testing and widespread use of the ISO-RILEM R679 mortar strength test is encountered in the literature. In the 'spirit of membership of the European Union', the mortar prism test method of EN 196 was adopted in BS 12.

The compressive strength developed in a mortar (or a concrete) depends on the materials used, the mix proportions, the procedure employed in mixing and the efficiency with which the mix is compacted into a mould, as well as the temperature, humidity and time of curing. Materials, procedures and equipment to be used are, therefore, described in detail in standards. The sand specified in EN 196 is not limited to that extracted from one source, but it must be CEN certified and give results equivalent to those obtained with a *reference sand* which is defined as consisting of rounded particles in five specified size ranges between 0.08 and 2 mm, with $99 \pm 1\%$ retained on 0.08 mm sieve. In addition, the silica content must be not less than 98% and moisture no greater than 0.2%. Samples of the reference sand may be obtained by contacting the German Standards Institute (DIN).

The mortar mix specified is 3:1:0.5 by weight of sand, cement and water, respectively. It is cast into $40 \times 40 \times 160$ mm moulds. Flatness of the resulting surfaces of the mortar prisms is important because surface irregularities would concentrate stresses during compressive strength measurement and affect the result. Compressive strength may be related to the volumes of cement (C), water (W) and air (A) in a mix by Feret's empirical law (1896):

$$S = k \left(\frac{C}{C + W + A} \right)^2 \tag{6.1}$$

where k is a constant for the aggregates, cement and curing employed.

The volume of air present depends on the degree of compaction achieved and the object is to achieve full compaction using, in the EN 196 procedure, a jolting table or a vibrating table giving equivalent results. Each prism then contains approximately 600 g of mortar. Excessive compaction must be avoided, however, as it causes particle segregation. Some air may be entrained during mixing of the mortar and checks against a reference sand are important because some sands cause more entrainment than others, possibly because of an abnormal, although small, amount of clay and/or organic matter adhering to the grains.

The effect of the amount of water used is marked but easily controlled. Curing is carried out at $20° \pm 1°$, in a mist room for 24 h and then, after demoulding, under water. Compressive strength is usually measured after 2 and 28 days. The latter gives what is referred to as the *standard strength* and BS 12 classifies cements on the basis of the level attained (Table 6.1). For low strength classes the 2-day test is replaced by one at 7 days. After curing, a prism is superficially dried and tested immediately. It is first broken in flexure in a specified manner and then the separate halves broken in compression across the 40 mm thickness. Prisms are cured in batches of three yielding six results for compressive strength at each age. If any one result deviates by more than 10% from the mean it is rejected and if any one of the remainder deviates by more than 10% from the new mean, then all the results must be rejected and the test repeated.

Precision estimates in EN 196 are given as coefficients of variation for 28-day strength only. For reproducibility 'between well-experienced laboratories', a CV of 'less than 6% may be expected'. For repeatability 'within a well-experienced laboratory', a CV 'may be expected to lie between 1% and 3%'. Taylor (1994) pointed out that, expressed as 95%

confidence limits, these CVs imply $\pm 10.1\,\text{N/mm}^2$ and $\pm 5.0\,\text{N/mm}^2$, respectively, for a mean 28-day strength of $60\,\text{N/mm}^2$. Continuing cooperative testing between laboratories is expected to improve precision.

Compressive strength testing of cements is undertaken primarily to demonstrate the quality and consistency of the product. In addition, it gives the user some limited information on its likely performance in a 'production' concrete. Harrison (1990) gave a formula correlating EN 196 mortar prism and BS 4550 concrete cube compressive strengths for the same cement:

$$\ln(p/c) = 0.28/d + 0.25 \qquad (6.2)$$

where p is the mortar prism compressive strength (N/mm^2), c the concrete cube compressive strength (N/mm^2) and d the curing period in days at test. At 28 days this is equivalent to a strength ratio p/c of 1.30.

Some results for concretes prepared using the sand, granite and the procedure specified in BS 4550: Part 3: 1978 are given in Fig. 6.2. They illustrate the effect of cement surface area and that of water/cement ratio. In construction contracts, test cubes are prepared from samples of production concrete taken as it is placed. Since 28-day strength is regarded as an important indication of concrete quality for structural engineering purposes, various accelerated curing procedures involving elevated temperatures have been proposed to reduce the time needed to get an indication of its potential value. Unfortunately, correlations with strength developed under normal curing conditions are poor, presumably because at elevated temperatures there is a coarsening of the pore size

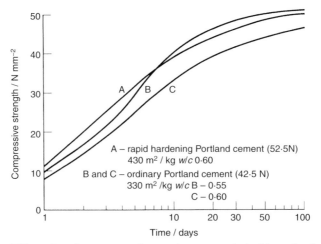

Fig. 6.2.   Effect of surface area and water/cement ratio (w/c) on the development of strength in BS 4550 concrete mixes (the w/c specified in the standard is 0.60)

distribution in the hydration products of the cement (Section 8.2.3). A prediction of 28-day strength from the 7-day strength, using a knowledge of the form of the growth curve, is preferred. This assumes that curves such as those in Fig. 6.2 may be displaced parallel to the strength axis but retain their shape, an assumption which is acceptable as long as the chemistry of the cement employed does not change significantly.

6.5 Workability

The measure of the ease with which a mortar or concrete can be placed and compacted is termed its workability and it can be assessed in a number of ways. Although there is no specification for workability in BS 12, it is an important property because good compaction of concrete is essential. The flow (rheological) behaviour of a mortar or concrete is usually dominated by the relative proportions of the aggregate, cement and water it contains. However, the hydration of a cement may cause premature stiffening and tests can be carried out with chosen aggregates to evaluate the contribution made by a cement to the workability of a mix.

Admixtures which modify flow are employed in mortars for masonry and increasingly in concrete (Section 10.1). Tests relevant to the behaviour of mortars in use (BS 4551, Part 1: 1998; ASTM C780) are not considered here. In current concrete practice, workability is measured in a number of ways which cover a range of conditions of use. Procedures are described in parts of BS 1881: Testing of Concrete. For example, *slump* (Part 102: 1983) is obtained by measuring the decrease in height under its own weight of a cone of concrete, while the *compacting factor* (Part 103: 1993) is obtained by measuring the degree of compaction achieved, as a relative bulk density, when a standard volume of loosely packed concrete is dropped from a fixed height into a cylinder. To test drier mixes which show little or no slump, the *Vebe test* (Part 104: 1983) is employed. In this test a cone of concrete is first allowed to slump normally and then vibrated to collapse to a given end point, the time for which is taken as an inverse measure of workability. For concrete of high or very high workability, the *flow test* (Part 105: 1984) is employed. It involves measurement of the spread of concrete, initially moulded as the frustum of a cone, across a steel plate which is hinged to a base plate at one edge and lifted at the opposite edge to a defined inclination and dropped. This is repeated 15 times in a period of 45–75 s. The uniformity of the spread concrete provides an indication of its cohesiveness.

Correlation between different workability tests is often poor because fresh cement paste is non-Newtonian and has flow properties which are influenced by the amount of shear it experiences, both in mixing and in the test measurement. In addition, changes in workability occur as the cement hydrates, although initial reaction soon subsides to a low level (Chapter 7). The problem with a fundamental approach to the rheology of concrete and mortar is that they contain aggregate particles which are too coarse for a conventional viscometer. Problems also arise from sedimentation of the solids and slippage at the surface at which shear is applied, as well as the possibility of shear inducing turbulent flow and air entrainment. Nevertheless, methods have been proposed which seek to overcome these difficulties and characterise concrete by determining its yield stress and plastic viscosity (Tattersall, 1991; de Larrard and Hu, 1994).

Cement influences the flow properties of a concrete or mortar by virtue of the smaller particle size range it introduces into the mix; but because it is reacting chemically with the water it also introduces a time dependent stiffening. The first of these factors is responsible for the plasticity and cohesion of the mix, the second can cause problems if the mix stiffens too much between production and placing and the probability of significant stiffening occurring is increased when mixing is of too short a duration. If the cement is seriously under-retarded because insufficient gypsum has been incorporated into it, then as soon as it is mixed with water a great deal of heat will be evolved and in extreme cases *flash set* will occur. The paste, concrete or mortar could not then be remixed satisfactorily. *False set* is different in that it can be removed by remixing; it is associated with the crystallisation of calcium sulfate dihydrate and sometimes referred to as *plaster set*. Flash set is extremely rare and false set is avoided by paying attention to cement milling temperature and the composition of the gypsum employed. The chemistry of these phenomena is considered in Section 7.3.

As the effects of cement hydration are time dependent, they require a test including time as a variable and that allowing the most straight-forward interpretation of the results is the ASTM False Set Mortar test (C359). Penetration of the mortar (water/cement ratio 0.3) by a 10 mm diameter plunger placed on its surface is measured at selected times and finally after remixing. An undisturbed area of mortar must be used for each measurement. The apparatus holding the plunger is similar to that for the Vicat needle and the total load on it is 400 g.

Results obtained in the false set mortar test are given in Fig. 6.3 and, while those for cement B would be considered normal, those for cement A provide an example of false set involving a marked induction time. The results for cement C might be considered symptomatic of a degree of under-retardation but only an examination of the hydration products can confirm this, because protracted crystallisation of gypsum can produce a similar result. It is important to remember that aeration of a normal cement will increase its tendency to cause mortar stiffening and if it is

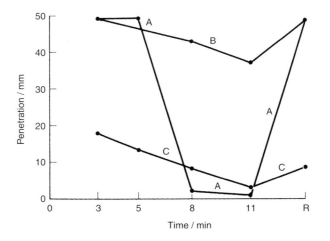

Fig. 6.3.   Results obtained in the ASTM False Set Mortar Test illustrating three types of behaviour (Section 6.5)

exposed for too long the cement may become false setting. Small samples must therefore be handled carefully. Information on the behaviour of a cement may be usefully enhanced by extending the time for which mortar stiffening is monitored and by introducing mixing time and water/cement ratio as variables (Bensted and Bye, 1986).

6.6   Soundness

A cement is said to be unsound if the hydration of a hardened paste of it is eventually accompanied by excessive expansion, causing cracking and reduction in strength. Accelerated hydration procedures are used routinely in cement production to check that it does not possess this property. The simple apparatus devised by Le Chatelier (Fig. 6.4) is used to indicate the expansion of a paste. In the EN 196 procedure the cylinder of paste of standard consistence is cured in a humidity cabinet ($\not< 98\%$ rh) for 24 h in the Le Chatelier apparatus, with glass plates covering the ends of the cylinder. The separation of the pointers is measured and the apparatus is immersed in water which is brought to boiling in 30 min and held there for 3 h. After cooling, the separation of the pointers is again measured and the change represents the expansion of the cement paste.

The limit for acceptable expansion is 10 mm (Table 6.1) and if this is equalled or exceeded then the cement is retested after a specified period of exposure to moist air. Expansion measured in the Le Chatelier test results from the delayed, slow hydration of hard burnt free lime crystals in the hardened paste. Aeration reduces expansion by converting some of the more reactive lime to calcium hydroxide and carbonate. Most UK

Fig. 6.4. Le Chatelier apparatus: the opening of the split brass cylinder by an expanding paste is magnified at the tips of the rods

cements produce an expansion of around 1 mm and in most countries where clinker chemistry and burning are well controlled, failure is extremely rare.

Expansion caused by the very slow hydration of magnesium oxide (periclase) crystals is not a problem in the UK where limestones low in magnesium are abundant. Where there is a potential problem however, accelerated curing in an autoclave is necessary for its detection. The expansion observed is dependent on the size as well as the quantity of periclase crystals present. Rapid cooling of a clinker in production reduces the expansion observed in the autoclave test by limiting periclase crystal growth. Free lime contributes to the measured expansion which also increases with an increasing content of C_3A in the cement. In the ASTM procedure (C151) mortar bars ($1 \times 1 \times 11$ inches) are cured at 216° in saturated steam under pressure for 3 h. The expansion observed must not exceed 0.8%, but such a specification can only be regarded as a general indication of the limit of acceptability because the microstructure produced at 216° differs considerably from that found at ordinary temperatures.

Another factor causing long-term expansion is an excessive level of sulfate in the cement. It is not certain to what extent this is detected by the above tests, but upper limits based on experience (expressed as $\%SO_3$) are given in all cement standards. The long-term effect of an excessive sulfate level is said to stem from the delayed crystallisation of sulfoaluminates (Section 7.4). BS 12: 1996 specifies a maximum sulfate content of 3.5% for all strength classes.

There are two theories of the mechanism of these undesirable expansions (Hansen, 1968). One is based on the idea that a swelling pressure is generated by the formation of colloidal reaction products which, because of their microporous nature, imbibe water with consequent swelling; the other on the idea that in the chemical reaction, formation of the crystals of the new phase generates a crystal growth pressure. The molar volumes in Equation (6.3) make it clear that overall volume change in a reaction is not necessary for expansion to occur.

$$CaO + H_2O = Ca(OH)_2 \qquad (6.3)$$

Molar volume/cm^3 17.0   18.0   33.1

Two factors may offset the shrinkage indicated in Equation (6.3) and contribute to expansion. First, the products of the reaction may be more porous than the reacting solid. In its reaction with water, hard burned lime will possess little porosity but the hydroxide can develop in the form of polycrystalline aggregates with significant porosity. Second, the water involved in the reaction may not be present at the reaction site if the lime is located in a dense, polymineralic grain of slowly reacting clinker phases. Water for the reaction must then either diffuse in from outside the hardened paste or from more porous, water-containing regions within it. As a result of the increase in solids volume accompanying the reaction, a significant localised swelling pressure would then be generated with a potential for cracking.

The driving force in the crystal growth pressure mechanism of swelling is the supersaturation of the solution of the product phase and it increases as the supersaturation increases. However, swelling does not automatically result from an increase in solids volume (Taylor, 1994). If the supersaturation remains low, crystals of the product may grow slowly and be accommodated in the pore system of the matrix. When this is not possible, the reaction may be stopped if the matrix is strong enough. If a paste is still plastic enough when an expansive reaction occurs, it can relax to reduce the internal stress. Only the delayed formation of sulfoaluminates or the hydroxides of lime or magnesia are associated with undesirable swelling. Where cracking does occur, enhanced transport of reaction product away from the reaction site will result in a change in the relationship between degree of reaction and swelling.

6.7   Heat of hydration

Heat of hydration is not determined as a routine test measurement but it is of major importance when a cement is being supplied for use in a large concrete structure with a low surface to volume ratio. In such constructions a high rate of heat evolution in the early stages of cement hydration, producing too big a temperature differential between the surface and core of the structure, must be avoided or stresses sufficient to cause cracking may develop. Temperature rises of several tens of degrees are recorded after three days in concrete cured adiabatically, the actual rise depending on the composition, fineness and proportion of the cement employed.

For prolonged curing periods, a heat of hydration can be derived from the heats of solution (ΔH^{sol}) of anhydrous and hydrated cement. The quantities involved are related by a thermodynamic cycle:

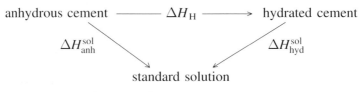

so that $\Delta H_H = \Delta H^{sol}_{anh} - \Delta H^{sol}_{hyd}$.

In the procedure given in EN 196 (Part 8 currently in draft) a nitric/hydrofluoric acid mixture is used to dissolve the fresh cement and a sample hydrated at $20 \pm 0.2°$, each containing the same weight of the former. The difference in the concentrations of the solution resulting from the water in the hydrated sample is negligible. The loss on ignition of a fresh cement is determined and the heat of hydration ΔH_H for the chosen curing period related to the 'loss free' mass of cement. The temperature rise accompanying dissolution of a sample is measured with a Beckmann thermometer or an alternative reading to $0.002°$ and additional temperature readings are recorded before and after the actual dissolution, to obtain a cooling correction. The acid is held in a resin-coated Dewar flask to minimise heat losses. The heat capacity of the apparatus and acid together is determined using Analytical Reagent grade zinc oxide since the heat of solution of this substance is accurately known.

It must be noted that what are termed heats of solution and hydration in the industry are exothermic and quoted as positive, while the corresponding enthalpy changes are negative, that is heat is lost from the system. The chemical reactions involved in hydration do not have a precise stoichiometry nor is the cement examined fully hydrated. Furthermore, because the hydration products have a much higher surface area than the anhydrous cement, significant changes in surface enthalpy and the enthalpy of adsorption of water on the hydrated material are therefore involved in the reaction (Brunauer and Kantro, 1964). However, the quantity ΔH_H is relevant to concrete practice.

EN 196 (Part 9 currently in draft) includes a method of determining heat of hydration 'during the first few days' (typically two days) using a semi-adiabatic (Langavant) calorimeter. In this test a mortar having the same composition as that used in the compressive strength measurement is held in a disposable cylindrical box inside a Dewar flask. A year-old mortar is held at room temperature in a reference calorimeter and the temperature rise in the freshly prepared mortar is measured with a platinum resistance thermometer to within $0.3°$. Both the total heat of

Table 6.2.   Total heats of hydration of individual cement compounds

Compound	Products	J/g*	Reference
C_3S	C–S–H, CH	(i) 460 ± 8	Fujii and Kondo (1976)
		(ii) 502	Lerch and Bogue (1934)
C_2S	C–S–H, CH	259	Lerch and Bogue (1934)
C_3A	C_3AH_6	866	Lerch and Bogue (1934)
C_3A[†]	$C_3A.3C\bar{S}.H_{32}$	1452	Lerch and Bogue (1934)
C_4AF[†]	$C_3(AF).3C\bar{S}.H_{31}$[‡]	724	Fukuhara et al. (1981)
CaO	CH	1168	Thorvaldson (1930)
$CSH_{0.5}$	CSH_2	125[§]	Bye and Coole (1984)

* Per g anhydrous compound.
[†] Hydrated in the presence of an excess of calcium sulfate — the water content of the sulfoaluminate is somewhat variable.
[‡] A/F probably 3 with hydrous iron oxide as a second product.
[§] Determined by conduction calorimetry — this increased to 170–180 J/g of starting material as the water content fell to around 0.05 H_2O.

hydration to chosen times (J/g) and the corresponding instantaneous rates of heat evolution (W/kg) are determined. It has been established that the heat of hydration obtained by this method after 41 hours corresponds to that obtained by the heat of solution method for a sample hydrated for seven days. The semi-adiabatic procedure provides a more realistic temperature profile for the hydrating specimen than isothermal methods. This is especially important when comparing cements which have different responses to temperature increase (Chapter 9).

Total heats of hydration for the principal compounds present in cement have been determined and values are given in Table 6.2. The large contribution made to cement hydration by C_3S is immediately apparent and C_3A, although present at a much lower level, also contributes significantly to early heat evolution because about 60% of it hydrates in the first day. If 50% of the C_3S hydrates in the same period, the heat liberated from these two phases will be of the order of 180 kJ/kg for ordinary (42.5 N) Portland cement.

. The British standard for *low heat* Portland cement (BS 1370: 1979) specifies a maximum lime saturation factor of 0.88 to limit the content of C_3S. It also specifies a minimum specific surface of 275 m^2/kg, since segregation of the concrete with water rising to the surface (bleeding) would result if reduction in heat evolution was obtained simply by reducing cement fineness. The limits specified are 250 kJ/kg at 7 days and 290 kJ/kg at 28 days (heat of solution method). The low early reactivity results in relatively slow strength generation, although ultimate strength is unimpaired. Low heat cement is not currently manufactured in the UK but the requirement can be met by using a blended cement (Chapter 9).

The heats of hydration for a range of calcium sulfate 'hemihydrate' compositions (Table 6.2) were determined with a conduction calorimeter. In this calorimeter, the heat produced by the hydrating compound (or a cement) is continuously conducted from a metal container holding the sample into a thermostat bath. Since the temperature difference between the two is never more than about 0.1°, the method is essentially isothermal. A thermopile is placed between the sample container and a large block of anodised aluminium which is in contact with water circulated from the thermostat. The e.m.f. generated in the thermopile by the small temperature gradient, which changes as hydration proceeds, is converted to a plot of heat liberation rate as a function of time (Fig. 7.1). The equipment is calibrated electrically using a resistor, inserted next to the sample holder, after heat liberation from a sample has fallen to a negligible level.

6.8   Concluding remarks — durability of concrete

National and international standards are designed to protect the purchaser of Portland cement by specifying requirements for those properties influencing its performance. However, while the tests described in this chapter focus on ensuring that a cement can be used to produce a concrete with the required strength, they cannot ensure that the concrete will possess adequate durability. This property of a concrete will be dependent on its *permeability* and it is important to realise that a concrete may reach the strength required for a particular structural purpose, even though its permeability may be unacceptably high. For example, the importance of a cover of dense concrete with a sufficient depth to inhibit rusting of either steel reinforcement rods or prestressing wires cannot be over-emphasised. Dense concrete will limit carbonation, which reduces the passivating effect of the cement, and inhibit corrosion of the steel by resisting the penetration of water, oxygen and chloride ions. There are situations, such as building foundations in a sulfate-rich environment, where a particular cement composition is required to resist attack by an aggressive agent, but a concrete with a low permeability remains important (Section 9.2). Factors influencing concrete durability are discussed in BS 5328 Part 1: 1997 — Guide to specifying concrete. Guidance on specifying durable concrete which may be subjected to carbonation and chloride-induced corrosion of reinforcement, freeze-thaw conditions, or chemical attack, has also been recently provided (Hobbs, 1998).

The ultimate permeability of compacted, hardened cement paste is very low (Section 8.3.1) and the permeability of concrete is determined by the efficiency with which the cement hydration products fill the voids between the grains of aggregate. Causes of high permeability in a concrete are:

(a) cement content too low;
(b) poor dispersion of cement in the mix;
(c) poor compaction when the concrete is placed;
(d) premature drying so that the degree of hydration of the cement is too low;
(e) too much water in the mix.

Cause (d) may result in cracking which persists in spite of further hydration in use and (e) will produce a concrete with an increased proportion of larger pores, to which excessive drying shrinkage will contribute. Any cracking accompanying such shrinkage will considerably increase permeability. An excess of water over that needed to hydrate the cement is necessary to make proper compaction possible but the excess must not be too great. The use of water as a means of achieving good workability is dangerous. An adequate cement content and its efficient dispersion in the concrete not only improve workability but also have a beneficial effect on permeability. Admixtures which enhance workability and make it possible to use a reduced water content are increasingly employed where high quality concrete is required (Section 10.1).

7. The hydration of Portland cement

7.1 Introduction

7.1.1 General introduction

When Portland cement reacts with water, heat is evolved and with a sensitive conduction calorimeter (Section 6.7) the rate of heat liberation can be followed while hydration proceeds almost isothermally. An example is given in Fig. 7.1 and the complexity of the hydration process is immediately apparent from the three peaks in heat liberation rate, which suggest that there are three maxima in the rates of the hydration reactions.

The first peak (I) is by far the highest but of short duration and followed by a dormant or induction period in which the heat liberation rate is low, although never zero. The physical changes in the paste during this period are readily detected by an increasing stiffness which can be quantified by means of a penetrometer (Section 6.5). The chemical and physical processes occurring are therefore of practical importance since they lead to a decrease in workability and the time limit on placing concrete is reached about halfway through the 'chemical' dormant period at normal water/cement ratios.

The dormant period ends with an acceleration in heat evolution during which setting occurs. In Fig. 7.1 the time-scale is such that changes in slope are considerable. Since the heat evolution rate usually changes rather slowly towards the end of the dormant period this may have to be defined somewhat arbitrarily, in comparing cements, for example. Most ordinary Portland cements when hydrated at 20° give a broad peak (II) similar to that in Fig. 7.1, with a maximum at 9–10 h. Not all cements produce a third peak (III) and its height, breadth and the time at which it appears vary considerably with the composition and grinding temperature of the cement. It normally only appears as a shoulder on the broad peak II with cements having the more usual C_3A contents of 8–10%. A small fourth peak has occasionally been observed. After three days, heat

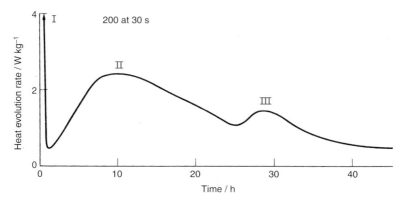

Fig. 7.1.   *Variations in the rate of heat evolution in the hydration at 20°C of an ordinary (42.5N) Portland cement containing 13% C_3A (Bogue) at a water/ cement ratio of 0.4*

evolution rates are low and the heat of hydration determined by the heat of solution method (Section 6.7) is usually employed to monitor further hydration.

To understand the reactions occurring in the hydration of cement and to link them with setting and the development of strength in the hardened paste, we need to know:

(*a*) how the hydration reactions of the individual compounds in cement contribute to the heat peaks in Fig. 7.1;
(*b*) the causes of the considerable changes in heat evolution rate;
(*c*) how the products of the hydration reactions pack together to fill space;
(*d*) the nature of the bonds between the hydration products in the hardened paste.

The engineering properties of concretes and mortars, determined by these chemical and physical aspects of cement hydration, include those which can be evaluated in the performance tests described in Chapter 6 and also permeability and durability, which are related. Chemical aspects ((*a*) and (*b*)) are considered in this chapter and physical aspects in Chapter 8.

An alternative to this physico-chemical approach involves the determination of empirical relationships between the results obtained with a cement in performance tests, its constitution and the conditions it has experienced in manufacture, without consideration of mechanisms of hydration. However, statistical correlations are often limited in applicability because too few results can be obtained for adequate treatment of the large number of variables which must be considered if useful generalisations are to be made. Sometimes the spread of values for a compositional or a process variable in such a study is too small for its significance to be identified by statistical techniques, especially if there is no foreknowledge of its existence.

Knöfel (1979) examined the empirical method to establish its limitations and found that, although the correlation between strength generation and major phase composition was good for experimental cements prepared from pure samples of the four main oxides, as soon as minor components and variation in pyroprocessing were introduced, correlations degenerated into 'vague trends'. Poor correlations were also obtained for commercial cements, unless limited to samples from a single works in which the content of minor components was approximately constant.

Historically, two mechanisms of hydration have provided the basis for the interpretation of experimental observations. Le Chatelier (1882) considered that cement hydration occurred by the dissolution of the anhydrous phases followed by the crystallisation of the hydrates in an interlocking mass. Michaelis (1893), on the other hand, thought that solidification of a paste occurred by the formation of colloidal material which hardened as it lost water to hydrate further anhydrous material. It will be seen that there are elements in both hypotheses that are still relevant.

Major factors determining the rate and mechanisms of cement hydration were initially identified in investigations of the hydration of laboratory-synthesised samples of the pure individual compounds present in it. This approach made a significant contribution but modern knowledge of the complexity of cement clinker composition and microstructure has revealed its limitations. Application of the range of techniques available in electron microscopy to the characterisation of the microstructure of clinker particles and of reaction products in the early stages of hydration has greatly aided interpretation. Backscattered electron imaging and improved precision in quantitative X-ray diffraction analysis (Chapter 4) have proved valuable in following the progress of hydration with curing time. Solid state nuclear magnetic resonance is introduced in this chapter. It has significantly advanced the characterisation of poorly crystalline phases, such as calcium silicate hydrate, and has also been used in monitoring the hydration of individual clinker phases in a cement.

7.1.2   Methods of investigating the kinetics of cement hydration

A number of methods can be used to examine the rate of the hydration of cement or its constituent compounds separately and the results make possible the interpretation of the variations in heat evolution rates in Fig. 7.1. The usual method of measuring the rate of a chemical reaction involving a solid requires a determination of fractions reacted or formed

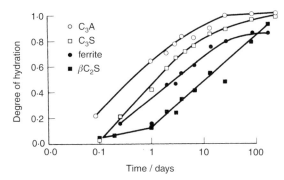

Fig. 7.2.   Degree of hydration at ordinary temperatures of the principal phases in a sample of ordinary Portland cement determined by QXDA (Copeland and Kantro, 1964)

as a function of time and it is possible to follow the consumption of each of the principal anhydrous phases in a cement by QXDA (Fig. 7.2). Dalziel and Gutteridge (1986) pointed out that for small amounts of reaction of a minor phase (such as ferrite at an initial 4% in their sample) the usual error in a determination of the unhydrated phase remaining in a sample results in a considerable uncertainty in the fraction hydrated. Consumption of anhydrous phases can also be followed quantitatively by backscattered electron imaging of hardened pastes in the scanning electron microscope and by nuclear magnetic resonance spectroscopy, considered in Sections 4.5 and 7.2 respectively.

X-ray diffraction is of limited value in examining the hydration products themselves because variations in their crystallinity as well as their quantity significantly influence diffracted intensity. However, either the chemically bound water in the hydration products, or the *non-evaporable* water after specified outgassing, is frequently determined as an indication of the degree of hydration. These are not absolute measures because neither the composition of the major hydration product (calcium silicate hydrate), nor the stoichiometry of the reaction, is well defined and also because some of the chemically bound water is less strongly held than some of the adsorbed water in the porous silicate hydrate. Consequently, any drying procedure which precedes the determination of loss on ignition (at 900–1000°) must be specified precisely when quoting results, since it influences the subsequent loss in weight.

The most common methods of drying hydration products involve displacement of water with a solvent, such as acetone or an alcohol, to stop hydration followed by evaporation of the solvent. However, both strong retention of the solvent and chemical interaction of it with cement hydration products have been observed. The use of propan 2-ol (isopropyl

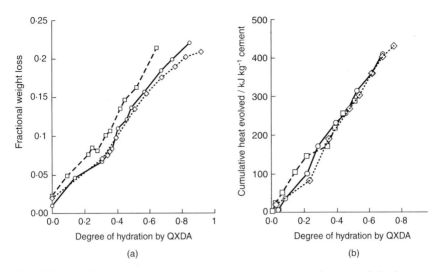

Fig. 7.3. Relationships for three cements between degree of hydration determined by QXDA and (a) fractional increase in bound water (weight loss 100–750° kg/kg anhydrous cement); (b) cumulative heat evolved (Parrott et al., 1990)

alcohol) followed by vacuum outgassing has been shown not to affect silicate anion structure formed in C_3S hydration (Brough *et al.*, 1994). If heating is used, temperature and time must be limited to minimise decomposition of hydrates. In any procedure the sample must be rigorously protected from atmospheric moisture and carbon dioxide, since their absorption complicates thermal analysis. The effect is much greater in small samples with a high surface to volume ratio than it is in bulk concrete and this is also true of the effect of CO_2 on the hydration reactions themselves.

In spite of these complications, a number of authors have used the increase in bound water to follow hydration, with reproducible results for both Portland cement and its constituent phases. A relationship with the degree of hydration of a cement determined by QXDA has been demonstrated and an example is given in Fig. 7.3(*a*). The slope of such plots depends on the composition of the cement and is sensitive to the experimental procedures used. A better correlation has been found for cumulative heat evolved and degree of hydration (Fig. 7.3(*b*)) and the conduction calorimeter provides a useful method of comparing the early reactivity of cements. It should be noted that the rates of early reactions are influenced by the mixing procedure used in 'wetting out' the solid cement to produce a paste. Samples should be gently rotated end-over-end before setting to prevent segregation.

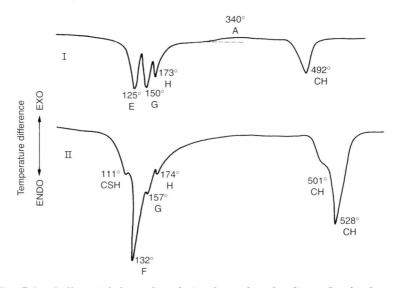

Fig. 7.4. *Differential thermal analysis of samples of ordinary Portland cement paste hydrated for 5 min (I) and 16 h (II). Endotherms for the dehydration of: E — ettringite; G — gypsum; H — calcium sulfate hemihydrate; CH — calcium hydroxide; CSH — calcium silicate hydrates; A — weak exotherm for the crystallisation of anhydrite (CaSO₄). The temperatures found for the peaks are particularly dependent on the sample (size, packing and composition) and the equipment used*

A reaction in which water, which has a high molar volume, is incorporated in a solid phase is usually accompanied by an overall shrinkage. The hydration of lime is a simple example (Equation 6.3). If a paste of cement is sealed at the chosen water/cement ratio, a *chemical (autogenous) shrinkage* is observed which can be measured with a simple dilatometer. The change with time can be recorded continuously, making it a convenient method of following the progress of hydration. Parrott *et al.* (1990) demonstrated a linear correlation of shrinkage with degree of hydration determined by QXDA; at 80% hydration the shrinkage measured was of the order of 50–60 ml/kg of cement.

More information concerning the water bound in cement hydration products can be obtained from differential thermal analysis, differential scanning calorimetry or derivative thermogravimetry. With careful standardisation of technique, peak areas in such plots can be used to follow hydration semi-quantitatively. These methods are also limited by the ill-defined stoichiometry of the hydration reactions and also increasing peak overlap after setting has occurred (Fig. 7.4). However, they are capable of discriminating between that calcium hydroxide formed by the hydration of tricalcium silicate and that formed by the

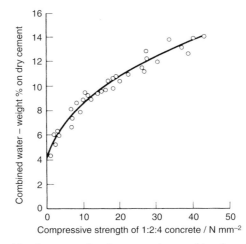

Fig. 7.5. Relationship between the increase in combined water content of cement pastes and the strength of identically cured concrete samples (Lea and Jones, 1935)

hydration of free lime, because this process produces poorly developed crystals of the hydroxide which have a lower decomposition temperature.

The CH peak is least affected by the long tail of the peak for calcium silicate hydrate and the determination of CH by thermogravimetry has been widely used to follow hydration of the silicates (Equation (7.1)) either alone or in Portland cement. Good correlations with the consumption of the anhydrous phases of a cement measured by the more quantitative methods have been reported, but the precise relationship will depend on the composition of the sample examined.

Determination of the development of compressive strength assesses hydraulicity rather than chemical reactivity. Nevertheless, the expected correlation between an increase in the strength of a concrete or mortar and an increase in the degree of hydration of the cement is observed. The non-linearity of the relationship shown in Fig. 7.5 can be ascribed to the importance to strength of a dense packing of hydration products, i.e. a low porosity. At a high degree of hydration an increase of 1% in bound water contributes more to the reduction of porosity than the same increase at a low degree of hydration, where the paste fraction of the concrete still has a relatively open texture (Section 8.3.2).

In addition to the changes in the amounts of the solid phases in a hydrating cement paste, there are significant changes in the concentration of dissolved species in the aqueous phase (Glasser, 1992). They can be followed by filtration, using pressure at later stages of hydration, and chemical analysis of the filtrate or expressed liquid, which must be free of colloidal material. Satisfactory protection against the absorption of CO_2 is

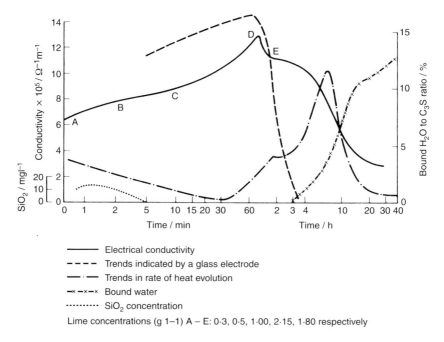

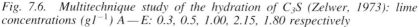

Fig. 7.6. Multitechnique study of the hydration of C₃S (Zelwer, 1973): lime concentrations (gl⁻¹) A—E: 0.3, 0.5, 1.00, 2.15, 1.80 respectively

only provided by manipulation of samples in an inert atmosphere. If a glass electrode suited to a highly alkaline medium is protected by a sheath of a semi-permeable membrane, to avoid damage to the glass and contamination of the liquid junction, then pH changes in the paste itself can be followed continuously. However, the potential generated by the electrode is determined by the hydrogen ion *activity*, and conversion of this to a concentration requires data on its activity coefficient in the presence of significant concentrations of potassium, sodium and calcium ions, which at elevated pH form ion pairs such as $CaOH^+$. Hydroxyl ion concentration is usually determined by titration of a filtrate with hydrochloric acid.

Results obtained by Zelwer (1973) for the hydration of C_3S, using a number of techniques, are given in Fig. 7.6 and they suggest a relationship between the maxima in pH, electrical conductivity and an acceleration in rate of heat evolution. Electrical conductivity measurements provide a simple method of following changes in the aqueous phase of a paste of a pure silicate, but with a commercial Portland cement a continuous increase in the concentration of dissolved alkalis distorts the curve and may even mask the maximum (D in Fig. 7.6).

7.2   Hydration of the individual phases in Portland cement

In this account of the mechanisms and rates of hydration of the four principal phases of Portland cement, only limited reference will be made to the refinements introduced by polymorphism and solid solution. Emphasis will be placed on the primary factors, although in the case of C_2S one of these is polymorphism (Section 1.3.2). Inevitably emphasis will also be on tricalcium silicate, the principal strength generating phase in cement. It has received the most attention in hydration mechanism studies and has been considered to provide a simple model for cement itself.

7.2.1   Tricalcium silicate

The products of the hydration of this phase at ambient temperatures are an ill-defined calcium silicate hydrate and crystalline calcium hydroxide (portlandite). The equation may be written:

$$C_3S + (y + z)H \longrightarrow C_xSH_y + zCH \tag{7.1}$$

$x + z = 3$ but x, y and z are not necessarily integers.

The calcium silicate hydrate formed in this reaction is poorly crystalline, yielding only two bands in its X-ray diffraction pattern, a broad one peaking at 0.305 nm and a weaker, sharper one with a peak at 0.182 nm. This and its variable composition have resulted in the use of the notation C–S–H to represent it, where the hyphens emphasise the indefinite composition. Because C–S–H has a high specific surface area, it has been described as a gel. It is important to note that the notation C–S–H is not used only for the hydration products of the calcium silicates in cement but also for poorly crystalline calcium silicate hydrates with a wide range of compositions, including those made from hydrated lime and colloidal silica, for example.

Determination of x, the C/S (= Ca/Si) ratio, is usually made by electron probe microanalysis or analytical electron microscopy (Section 4.5.2). A wide range of values has been reported for C_3S pastes but a generally accepted mean is 1.7–1.8. It is important to note that the existence of conflicting results in the literature may be due to differences in the probe size employed since the volume of the sample exposed to the electron beam may limit discrimination between the phases present. However, there appears to be a real variability in C–S–H formed from C_3S. Richardson and Groves (1993) examined ion beam thinned samples by transmission electron microscopy (TEM). They determined C/S ratios of C_3S pastes hydrated for extended periods, including one sample cured for 26 years by the Portland Cement Association and found mean values of 1.7–1.8, but with individual values ranging from 1.2–2.1 on the sub-micrometre scale.

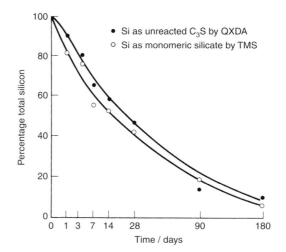

Fig. 7.7. *Relationship between two methods of following the hydration of C_3S (Dent-Glasser et al., 1978)*

The value of y is also difficult to specify because there is no distinct break in loss in weight between drying and dehydrating this compound. Drying, which is also a complication in determining the surface area of a paste (Section 8.2.1), has frequently been carried out at ambient temperature under vacuum, with the water vapour pressure maintained at that of ice at $-78°$ by means of a dry ice trap (Copeland, 1961). This corresponds to a pressure of 5×10^{-4} torr and since equilibrium is approached very slowly, drying is considered to have been completed when the rate of loss in weight of a small specimen falls to 0.1 mg per day (Brunauer, 1970). This procedure, which takes 4–7 days, is referred to as hard D-drying and it is only one of a number of arbitrary techniques which seek to minimise decomposition while drying samples of C–S–H. Brunauer and his co-workers obtained values close to 1 for y in D-dried pastes but values of 1.3–1.8 have been reported by later workers (Odler and Dorr, 1979). For fully saturated C–S–H, Young and Hansen (1987) found a value of 4.

When C_3S hydrates, the orthosilicate ions (isolated SiO_4 tetrahedra) are first converted to disilicate ions ($Si_2O_7^{6-}$) and gradually higher polymeric forms develop. The results in Fig. 7.7 show the relationship between the disappearance of anhydrous C_3S and that of the orthosilicate ions it contains. The degree of polymerisation of the silicate ions can be determined by trimethylsilylation (TMS) of hydrated pastes using a mixture of trimethylchlorosilane and hexamethyldisiloxane in dimethylformamide (Tamas, 1976). The first substance converts the less polymerised silicate ions in the specimen into volatile silyl esters, identified by gas-liquid chromatography using helium or nitrogen as a carrier gas. Non-volatile higher polymers can be identified by gel permeation chromatography. Results for C_3S and Portland cement pastes have shown that the dimer content reaches a maximum of around 50–60% after six months and then falls to around 40% after a year, while that of the polymeric forms steadily increases to 40–60% (Mohan and Taylor, 1982). The limitation of the TMS process is that side reactions can occur and only 80–90% of the silicon in a sample may be recovered, but it indicates the proportion of the silicon in the different polymers present.

An alternative way of examining silicate anion polymerisation is provided by nuclear magnetic resonance (NMR) spectroscopy, since the ^{29}Si isotope (natural abundance 4.7%) has an intrinsic magnetic moment, as do some isotopes of other elements important in cement chemistry, namely ^1H, ^{27}Al, and ^{17}O. Placing materials containing such isotopes in a magnetic field results in the splitting of otherwise degenerate energy levels to an extent dependent on the field strength. Resonant absorption of photons can then be observed if the sample is simultaneously exposed to pulses of radiation of an appropriate wavelength. At realisable magnetic field strengths (up to 14 Tesla), nuclear magnetic resonance spectra lie in the radio frequency region of the spectrum. One commercial spectrometer has a 4.7 T magnet and records spectra in the 40 MHz region. An introduction to the basic physics of NMR, including reference to factors which are important in obtaining quantitative spectra, is given by Granger (1994).

The fact that nuclear energy levels, and hence the resonance peaks observed, also depend on the immediate environment (local magnetic field) of the nucleus has resulted in the use of such spectra to probe the short range aspects of structure. This is of particular value with poorly crystalline materials such as C–S–H where the long range order required for X-ray diffraction is largely absent. In a solid, the environment of a nucleus is anisotropic, which results in considerable line broadening in the NMR spectrum and this initially limited its application mainly to the study of liquids, the molecules in which are freely rotating. However, by spinning a sample of a solid (at a frequency of several kHz) at the 'magic angle' of 54° 44′ to the magnetic field, the orientation effect is eliminated and resolution of the fine structure in a spectrum is possible (Andrews, 1981). The abbreviation MAS-NMR is usually employed in the literature to emphasise that this technique has been used. Resolution of peaks in spectra is impaired if the sample contains paramagnetic ions such as iron, chromium or manganese in solid solution in the phase containing the isotope being excited. For example, peaks for the different Si site environments in tricalcium silicate merge into a single broad one in alite (Fig. 7.8(a)).

The effect of its environment on the absorption by a nucleus is recorded as the displacement of the resonance frequency (ν_s) in relation to

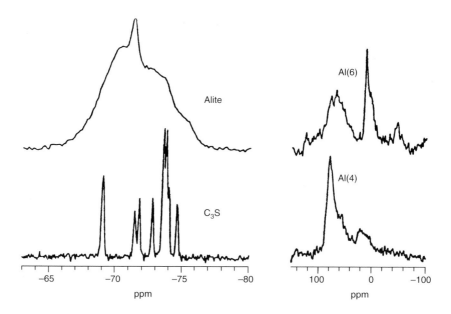

Fig. 7.8. *(a)* ^{29}Si *MAS NMR spectra of alite and* C_3S*; (b)* ^{27}Al *MAS NMR spectra of a sample of Portland cement before and after partial hydration, in which Al coordination changes from 4 to 6 (Kirkpatrick and Cong, 1993)*

that of a chosen standard (ν_{ref}), silicon tetramethyl in the case of ^{1}H and ^{29}Si, aluminium chloride or nitrate solutions in the case of ^{27}Al. The 'chemical shift' (ν_{s-} ν_{ref}) is recorded in ppm as the ratio of a shift in Hz to a resonant frequency on a scale of MHz. This eliminates the actual magnetic field strength employed and so makes results from different spectrometers comparable.

An obviously major alteration in the local environment of an atom results when its oxygen coordination number is changed, as is the case with aluminium in the hydration of the tricalcium aluminate in Portland cement (Fig. 7.8(b)). The nearest neighbour environment of silicon in cement compounds is almost invariably 4 coordination in the tetrahedral SiO$_4$ group. However, next nearest neighbour influences are changed on hydration, as a result of the sharing of oxygen atoms by adjacent SiO$_4$ tetrahedra. Chemical shifts have been established with the aid of crystalline silicates of known structure. Q with a superscript or a subscript is used to indicate the connectivity of an SiO$_4$ tetrahedron in a silicate anion (Table 7.1). Thus Q^0 is the range of shifts found for ^{29}Si in orthosilicates, Q^1 for disilicates or the end tetrahedra of a chain, Q^2 for middle SiO$_4$ in a chain, Q^3 for branching sites in a chain and Q^4 for a

Table 7.1. Typical NMR shifts for silicon in cement phases

	Designation	ppm
Ortho (mono) silicates	Q^0	−66 to −74
Disilicates and chain end groups	Q^1	−75 to −82
Mid-chain groups	Q^2	−85 to −89
Chain branching sites	Q^3	−95 to −100
Three-dimensional network	Q^4	−103 to −115

three-dimensional network in which every O in the SiO_4 tetrahedron is shared by two Si atoms, as in a quartz (SiO_2) crystal.

Only Q^1 and Q^2 resonances have been found in C–S–H and while TMS indicates the proportion of silicon in the different silicate anions present in a sample, NMR indicates a mean chain length from the ratio of end group absorption peak areas to those for mid-chain groups. NMR has the advantage that samples do not have to be dried nor does it involve chemical treatment which might produce side reactions. In following the hydration of C_3S, Clayden *et al.* (1984) found a good correlation between the degree of hydration indicated by the changing relative areas of Q^0 and Q^1 peaks, the increasing cumulative heat liberated and the increase in the amount of CH formed. Rodger *et al.* (1987) showed that the Q^1/Q^2 ratio reached a maximum after six months in a C_3S paste, corresponding to the maximum in dimer content indicated by TMS. Further hydration resulted in a slow, small decrease in this ratio indicating a higher degree of polymerisation. They found a ratio of 42 end units to 58 middle units in a sample of C_3S paste hydrated for 26 years, indicating a mean chain length of 4.8. To simplify interpretation of spectra, they used a cross-polarisation technique in which only Si atoms close to protons absorb radiation so that the resonances of the anhydrous phases are eliminated.

Brough *et al.* (1994) used MAS-NMR to follow the hydration of a C_3S paste sample held in the magnetic probe. Hydration was found to be little affected by the spinning. An improved signal to noise ratio and hence rate of recording of spectra was made possible by using C_3S containing ^{29}Si enriched to 100%. Samples were hydrated at 25°, 50° and 75° and as well as the expected increase in hydration rate, increasing temperature resulted in increased polymerisation at a given degree of hydration. They also observed that a small amount of hydrated monomer, previously observed in NMR spectra only in the dormant period, persisted throughout the hydration. From a consideration of possible methods of polymerisation, they concluded that disilicate ions formed initially were subsequently linked to form higher polymers by the hydrated monosilicate present. For example:

dimer + monomer + dimer $\longrightarrow$ pentamer

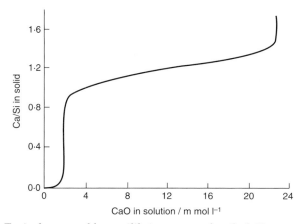

Fig. 7.9. Typical metastable equilibrium curve for C–S–H composition as a function of calcium hydroxide in solution expressed as CaO, after Steinour (1954) (courtesy of H.F.W. Taylor)

This is in accord with the 2, 5, 8 ... $3n - 1$ sequence of polymers reported by Hirljac *et al.* (1983).

A number of structural models have been proposed for C–S–H and a detailed discussion of them is given by Taylor (1997). Brunauer and Kantro (1964) argued that the X-ray diffraction bands given by the C–S–H formed from Portland cement could be indexed as lattice spacings of the mineral tobermorite (approximately $C_5S_6H_5$), which has a layer structure with anionic layers containing silicate chains. They introduced the term *tobermorite gel*. Taylor (1981) pointed out that the 0.31 and 0.18 nm spacings are close to prominent repeat distances within a layer of the portlandite structure and suggested that the small degree of longer range order must reside in the Ca–O parts of the gel structure.

C–S–H with a wide range of C/S ratios can be prepared by the interaction of lime with silicic acid, $Si(OH)_4$, or with very fine silica fume, in an aqueous suspension. Fig. 7.9 is typical of published results for the metastable 'equilibrium' of C–S–H samples and CH (expressed as lime) and hydrous silica in solution, indicating the significant step up necessary to inhibit leaching from material with a C/S ratio greater than about 1.3.

Cong and Kirkpatrick (1996) examined a range of preparations of this type with C/S from 0.44 to 1.9 using ^{29}Si and ^{17}O NMR spectroscopy. At low C/S, the silicate anion polymerisation was found to be extensive and a relationship to tobermorite appears meaningful (Fig. 7.10). The chains containing repeat units of three tetrahedra are referred to as *dreierketten* and some replacement of Si by Al can occur in the central bridging tetrahedra of the units if an aluminium containing phase is hydrating at the same time as C_3S. Taylor (1993) suggested that as C/S increases towards

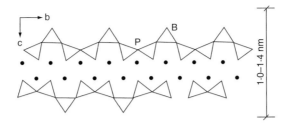

Fig. 7.10. Schematic representation of dreikette-type chains of SiO₄ tetrahedra (△) in a single layer of the tobermorite structure. The lattice 'c' spacing is reduced by strong drying to 1.0 nm. Ca ions (●) are coordinated to unshared oxygen atoms in the paired tetrahedra (P) which are linked by bridging tetrahedra (B). Missing bridge tetrahedra increase in number to accommodate more calcium ions in increasingly disordered higher C/S ratio forms of C–S–H. Interlayer calcium ions and water are omitted

the higher levels observed in hydrated silicates, the anionic layers in C–S–H consist of an increasing proportion of distorted jennite units intermixed with those of tobermorite. Jennite ($C_9S_6H_{11}$) is also a mineral with a layer structure. An increase in vacant SiO_4 bridging sites would also be expected as C/S increases and an examination of samples using ^{17}O and ^1H NMR has suggested that increasing association of calcium ions and hydroxyl ions and an increasing number of Si–O⁻ groups can be considered as part of the charge balance. However, at high C/S values, the number of dimers present is such that Grutzek *et al.* (1997) considered that the calcium silicate hydrate is then better described as a distinct phase in the C–S–H system, a calcium disilicate with the general formula $Ca_4Si_2O_7 (OH)_2 \cdot xH_2O$.

7.2.1.1 Kinetics of tricalcium silicate hydration.   The progress of the reaction of C_3S with water can readily be monitored in the conduction calorimeter, and Fig. 7.11 contains results demonstrating the effect of surface area, which is that to be expected for a reaction involving the interface between a solid and water. As with Portland cement there is an initial peak, followed by a dormant (or induction) period and then a major peak in heat evolution. Comparing Figs 7.11 and 7.1, it is seen that there is a correspondence between peak II for C_3S and that for cement, although for the C_3S sample with normal cement fineness the peak occurs somewhat later. However, the soluble sulfate and alkalis present in commercial Portland cement have been shown to accelerate the hydration of C_3S, and this phase (as alite) makes the major contribution to peak II in cement hydration.

The heat liberation curve is, of course, the derivative of the cumulative total heat curve. After a small initial step corresponding to peak I, the

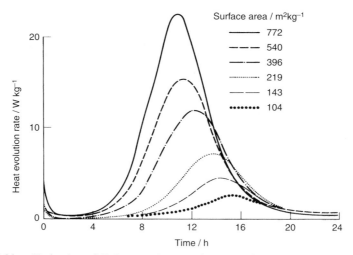

Fig. 7.11. Hydration of C₃S pastes in a conduction calorimetre: effect of surface area (Tenoutasse, 1969) (water/solid ratio = 1.0)

latter has the same sigmoidal form as the more usual plot of fraction reacted (α) against time (t) for a solid state reaction and the Avrami-Erofe'ev Equation (7.2) can often be fitted to a significant portion of such a curve:

$$-\ln(1 - \alpha) = kt^n \tag{7.2}$$

where k is a rate constant.

Using this equation with results for C_3S, Tenoutasse and de Donder (1973) obtained a value of three for n. In the thermal decomposition of solids, different values of n have been interpreted as indicating particular rate-controlling mechanisms but for a reaction in which water is reacting with a suspension of particles such mechanisms are not applicable. In a through-solution mechanism (Fig. 7.12(a)) growth rate at any instant might be expected to be dependent on the surface area of C–S–H already formed and Gartner and Gaidis (1989) found a good correlation between hydration rate and surface area at the acceleratory stage. Eventually a deceleration occurs, because of a build-up of a layer of hydration products around the hydrating particles. As in the oxidation of metals, the effectiveness of a layer of reaction product as a barrier to further reaction depends on its porosity and the rates of diffusion of ions in it.

Tenoutasse and de Donder obtained an apparent Arrhenius activation energy (E_A) of 34 kJ/mol for C_3S hydration and this might be considered too high for diffusion to be rate determining. However, decreasing values of E_A have been reported for Portland cement hydration as the degree of hydration (α) increases, suggesting a change in mechanism to one in which

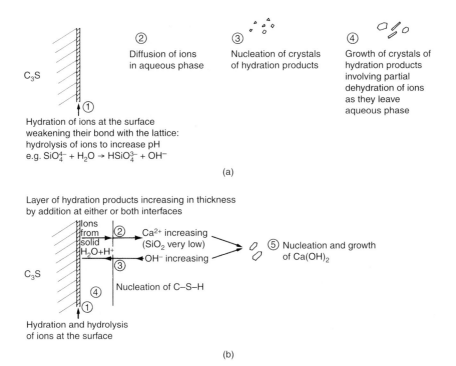

Fig. 7.12.   *Basic hydration mechanisms—possible rate-determining steps are numbered: (a) hydration through solution involving dissolution—diffusion—crystallisation;(b) hydration at the surface involving diffusion of ions through a product layer*

diffusion is rate determining. Hence there is the possibility that the value of E_A obtained in an investigation of hydration of a clinker phase may depend on the range of values of α covered by the experimental rate measurements. It should be remembered that the application of the Arrhenius equation to rate constants determined at several temperatures (T) assumes that the reaction involves a single rate-controlling step at all values of α and T examined.

Brough *et al.*, while conceding some limitation in their control of sample temperature, obtained significantly different apparent Arrhenius activation energies for dimer (35 kJ/mol) and higher polymer (100 kJ/mol) formation, which they were able to follow separately using ^{29}Si NMR with ^{29}Si enriched C_3S. The higher value for the latter correlates with the observation that the degree of polymerisation is greater at a given degree of hydration when an elevated curing temperature is employed.

Two classical mechanisms of hydration are shown schematically in Fig. 7.12 in which possible rate-determining steps are numbered. The 'through-solution' mechanism (applicable to gypsum plaster hydration, for example) is usually identified by the formation of a product at points remote from the dissolving phase. The solution may become supersaturated with respect to the hydration product and nucleation and growth of the latter will be rate determining if the rate of dissolution of the hydrating phase is high. In fact, the low solubility of C–S–H and the rapid establishment of a high supersaturation ensure that it is precipitated close to the C_3S surface as an amorphous product, so that distinction between the two mechanisms is difficult.

A topochemical reaction mechanism has been suggested in the past, but this implies a reaction in which the product is formed in the space originally occupied by a reactant, with a definable relationship between the crystal lattices of reactant and product. In view of the structural change involved in reaction 7.1, and not least because of the significant increase in total solids volume involved, a simple topotactic mechanism is not possible. Direct observation of hydrating systems using electron microscopy and an environmental cell has shown that, initially at least, a through-solution mechanism exists. Later in the hydration process, material is also progressively formed within the boundaries of anhydrous particles and designated as *inner product* (Sections 7.3 and 8.1).

The practical effect of a growing surface layer of hydration products is a long period of deceleration of reaction, to which the completion of the hydration of smaller particles makes a contribution. As Gartner and Gaidis (1989) pointed out, relatively little investigation of this stage of hydration has been made, although it is of great importance in determining the ultimate microstructure of a paste and hence strength development for engineering applications. At this stage, when the residual water content of a paste is low, a solid state reaction has been suggested in which diffusion through the product layer becomes rate determining. As indicated in Fig. 7.12, the most likely diffusing species maintaining a charge balance are H^+ to the reacting surface and Ca^{2+} and silicate species away from it. Any remaining alkaline pore solution within the product layer will greatly enhance their diffusion as well as increasing the mobility of silicate anions. The kinetics may be further complicated if separation (spalling) of hydrated material from the reacting surface occurs, which appears to be a possibility with C_2S (Section 7.2.2).

7.2.1.2 Origin of the dormant (induction) period. The existence of a period of low rate of reaction between two peaks is of interest because, although reactions involving solids often display periods of acceleration and deceleration, the existence of an acceleration after a period of deceleration is unusual. Because of its importance in the placing and proper compaction of concrete, the scientific basis of the dormant period and the mechanism by which it is terminated has attracted much discussion over a long period. Relevant facts concerning the early hydration of C_3S are summarised below.

(a) Immediately after contact of C_3S with water, both 'lime' and 'silica' rapidly appear in solution. Note that although analytical results are usually quoted for lime and silica, the species actually present in solution are calcium, hydroxyl and silicate ions. The concentration of dissolved as distinct from colloidally dispersed silica is significantly lower than that indicated in Fig. 7.6. A maximum concentration around $4.5 \, \mu mol/l$ $(0.27 \, mg/l)$ was observed by Brown et al. (1984) at a water/solid ratio of 0.7, falling to $0.5 \, \mu mol/l$ in the first 2 h of hydration. The proportion of C_3S hydrated in this time was 0.5% in an unstirred paste but nearly doubled by continuous stirring.

(b) It is generally accepted that a hydration product is rapidly formed as a layer on the surface of the C_3S particles at this stage and that it is responsible for the abrupt fall in heat evolution. It has been observed using HVSEM with an environmental cell to hold a paste (Sujata and Jennings, 1992; Meredith et al., 1995).

(c) The calcium ion concentration and pH in the aqueous phase increase to a maximum and a period of supersaturation follows (point D in Fig. 7.6). A very low rate of hydration persists throughout which has been shown to produce a hydrated monomeric silicate by ^{1}H NMR (Miljkovic et al., 1988) and ^{29}Si NMR (Rodger et al., 1987).

A number of authors have reported results suggesting a correlation between the end of the dormant period and the onset of the crystallisation of calcium hydroxide (e.g. Fujii and Kondo, 1975) but others have obtained contrary results. Seeding a paste with calcium hydroxide crystals in fact retards rather than accelerates hydration (Uchikawa, 1984). In an examination of the hydration of C_3S at a wide range of water/solid ratios, Damidot and Nonat (1991) found that the reaction rate accelerated before the precipitation of CH.

(d) When the surface area of the C_3S or the temperature of hydration are increased, the dormant period is shortened. Additions of colloidal silica (de Jong et al., 1967) and prehydrated C_3S (Odler and Dorr, 1979) both have an accelerating effect, while employing saturated CH solution as the hydration medium causes a significant lengthening of this period (Kondo and Daimon, 1969).

(e) The acceleration in heat evolution is accompanied by the formation of rapidly increasing amounts of C–S–H. The most frequently described forms are acicular and thin foils (Section 8.1).

(f) The reactivity of C_3S can be enhanced by rapid cooling after it is formed at high temperatures by sintering (Skalny and Maycock, 1974; Fierens and Verhaegen, 1975), thereby shortening the dormant period. Fierens and Verhaegen (1976) detected defects in the lattice of samples of C_3S using thermoluminescence spectroscopy and linked increasing defect concentration with increasing rate of cooling (between 1300° and 1100°) and decreasing length of the dormant period. They also found that this period could be significantly shortened by exposing C_3S to UV radiation which increased defect concentration.

In summary, two major factors influence the dormant period in C_3S hydration: its intrinsic reactivity and its interaction with dissolved lime. The latter may be that derived from its own initial hydrolysis or that added to the mix water (Nonat et al., 1997). However, the trigger mechanism terminating it remains elusive, especially since establishing correlations of events within the time scale is difficult. Potential complications in identifying a definitive correlation are: the effect of the water to solids ratio selected; the possibility that the C_3S sample contains a small amount of unreacted lime; the fact that the dormant period is short and its end can be difficult to define because acceleration in heat evolution is gradual; in a paste, local differences in aqueous phase composition may occur where agglomerates of anhydrous particles are not well dispersed.

Theories attempting to account for the dormant period and its termination, which have stimulated much discussion, are grouped below under four headings. More detailed accounts are given by Taylor (1997) and Odler (1998).

(i) Initial surface hydrolysis of C_3S (peak I) produces a low permeability layer on the surface of the grains and only when this layer is transformed into more stable, discrete particles of C–S–H does hydration accelerate.

(ii) An initial hydration product layer acts as a semi-permeable membrane behind which hydration proceeds as water passes through it (osmosis). The dormant period ends when the pressure build-up behind the membrane is sufficient to burst it and C–S–H grows outwards from the surface of the particles in the form of fibrils and tubules.

(iii) The dormant period is a classical induction period involving nucleation of C–S–H. It ends as the growth of nuclei gradually takes over as the rate-controlling process.

(iv) The dormant period results from the poisoning of calcium hydroxide nuclei by hydrous silica and ends when the supersaturation of the aqueous phase is high enough to overcome this effect and CH crystals begin to grow.

Suggestion (ii) is not now considered to be relevant except possibly at very high water/solids ratios and (iv) is considered to require a higher level of supersaturation than can be reached in solutions of calcium hydroxide. Microstructural studies have indicated the presence of a layer on the surface of C_3S particles in water and a convincing description of the dormant period is provided by theories (i) and (iii).

The barrier layer in the theories of group (i) has been variously described but the evidence indicates that it is a layer of metastable C–S–H containing hydrated monomeric silicate anions. Rodger *et al.* (1988) suggested that the end of the dormant period results from the development of conditions within it which favour the formation of dimers, the process which is predominant in the acceleratory period. Whatever the nature of the layer inhibiting reaction at the C_3S surface, solubility measurements have indicated that it appears to have an outer surface in metastable equilibrium with the aqueous phase (Jennings *et al.*, 1997). These authors suggested that the dormant period ends when the local supersaturation with respect to calcium within the metastable layer is sufficient to induce a 'eutectic-like' solidification of calcium hydroxide and a more stable C–S–H, no longer in contact with anhydrous C_3S. However, they noted that microstructural examination only revealed a change in morphology, and the solubility of C–S–H decreased, after the beginning of the acceleratory period at around the time of setting.

7.2.2 Dicalcium silicate

The hydration of dicalcium silicate, which is an orthosilicate, can only be studied if the hydraulic forms are stabilised by solid solution and, as pointed out in Section 1.3.2, reactivity and hydraulicity depend on the stabiliser and the amount of it used. The extremely slow rate of hydration of γC_2S has been said to be due to the fact that the arrangement of oxygen atoms around the Ca^{2+} ions in this form is a regular, octahedral one, but irregular in the other (hydraulic) polymorphs. This suggestion implies that thermodynamic factors, namely its lattice energy and the heats of hydration (solvation) of its constituent ions, are of primary importance in the hydrolysis of this silicate. However, it is more probable that rate factors are predominant in inhibiting hydration since that of γC_2S is almost as rapid as that of the β form if the pH of the aqueous phase is prevented from rising far by frequently replacing it with fresh water (Bogue and Lerch, 1927).

The products of the hydration of βC_2S are C–S–H and CH, the proportion of the latter being about one-fifth of that produced in the hydration of C_3S. Examination of hardened pastes by analytical electron microscopy has shown that the C/S ratios in the C–S–H formed lie in the range 1.6–2.0. The Q^1/Q^2 ratio in ^{29}Si NMR spectra indicated that the silicate chain length in older C_2S pastes was greater than that formed from C_3S, and that C_2S pastes are more ordered (Tong *et al.*, 1991). Cross-polarisation NMR revealed that, as with C_3S, early hydration led to the formation of hydrated monomeric silicate ions (Tong *et al.*, 1990, Cong and Kirkpatrick, 1993).

The hydration of βC_2S takes place much more slowly than that of C_3S (Fig. 7.2) and at first sight apparently contrary results have been obtained. However, Taylor (1997) pointed out that the variability in the hydration rates reported is likely to be due not only to the use of different stabilisers but also to the extent of ex-solution of material in solid solution at grain and lamellae boundaries as samples are cooled after firing. The finding that hydration can occur preferentially along the boundaries of twin planes (Scrivener, 1989) may be linked to this phenomenon.

With a sample stabilised by 0.5% B_2O_3, Tong and Young (1977) observed that the saturation ionic product of calcium hydroxide was only exceeded in the aqueous phase after 24 h. With a sample stabilised by 0.2% B_2O_3, Fujii and Kondo (1979) found a maximum supersaturation after about 80 h; the rate of hydration then increased to a maximum at 40 days (20% hydration) but unlike that of C_3S it exhibited a second acceleratory period after about 60 days. This acceleration suggests that spalling of a layer of hydration products had occurred at the surface of the unhydrated residual C_2S. These observations indicate that dicalcium silicate contributes little to peak II in the hydration of Portland cement but the results in Fig. 7.2 show that it will make a significant contribution to late strength if hydration is continued.

7.2.3   Tricalcium aluminate

Tricalcium aluminate reacts rapidly with water to form crystalline hydration products with different lime/alumina ratios:

$$2C_3A + 21H \longrightarrow C_4AH_{13} + C_2AH_8 \tag{7.3}$$

The products in this reaction are formed as platelets with hexagonal symmetry, their morphology resembling that of portlandite; they are metastable with respect to cubic C_3AH_6, to which they are rapidly converted at temperatures above 30°. However, the heat of hydration of C_3A is such that some C_3AH_6 may be produced rapidly, even in samples nominally hydrated at ambient temperature. If lime is present when C_3A is hydrating, as it is when C_3S is present, the formation of C_4AH_{13} is favoured, its conversion to C_3AH_6 is inhibited, and hydration proceeds more slowly. Nevertheless, reaction is sufficient, even at the level of C_3A

present in cement (commonly 8–10%), to induce flash set unless gypsum is added to act as a set regulator (Section 6.5).

In the presence of calcium sulfate, the product of hydration is a sulfo-aluminate, $C_3A.3C\bar{S}.H_{31-32}$, known as the mineral ettringite and usually referred to by this name. It is crystalline, forming prismatic or acicular crystals with a hexagonal cross-section. Solid solution in the structure is common (Fe^{3+} for Al^{3+} for example), and compounds with this structure are referred to as AFt phases. Substitution of sulfate by other anions such as chromate is also possible. Much of the water in the structure is loosely held and associated with calcium and sulfate ions in channels between columnar units which are parallel to the needle axis and which have the empirical formula $(Ca_3Al(OH)_6 \ 12H_2O)^{3+}$ (Moore and Taylor, 1968). Ettringite dissolves almost congruently in water at ordinary temperatures but, in the absence of lime and sulfate, increasingly incongruently as the temperature increases to form gypsum and aluminium hydroxide gel.

The retardation of C_3A hydration, in the presence of calcium and sulfate ions in the aqueous phase, has long been considered to be due to the formation of ettringite on its surface. Retardation is greater if calcium hydroxide is also present in solution (Forsén, 1938) and this is probably the result of the precipitation of a more compact layer of smaller crystals from a solution with a greater supersaturation. After an initial rapid reaction when a paste of C_3A and gypsum is mixed with water, there is a significant but declining level of reaction during what has been described as a semi-dormant period (Gartner and Gaidis, 1991). If there is insufficient calcium sulfate for complete conversion of C_3A to ettringite (Equation (7.4)), this period is followed by a sudden marked acceleration in reaction rate, at a time determined by the $\bar{S}/A$ ratio in the paste as well as the specific surface area of the aluminate.

$$C_3A + 3C\bar{S}H_2 + 26H \longrightarrow C_3A.3C\bar{S}H_{32} \qquad (7.4)$$

Changes in reaction rate are readily observed with a conduction calorimeter, and Tenoutasse (1969) showed that a second peak occurred when the gypsum in a paste had been consumed (Fig. 7.13). The magnitude of the initial peak, in which several percent of the C_3A may react, will depend on the reactivity of the latter as well as the rate of solution of the calcium sulfate. It should be noted that all, or almost all, of the calcium sulfate dihydrate (gypsum) used in the manufacture of cement is converted in milling to the 'hemihydrate'. Calcium sulfate hemihydrate dissolves much more rapidly than the mineral gypsum and establishes a higher supersaturation with respect to the sulfoaluminate. These properties of the hemihydrate and the simultaneous formation of CH by the hydrolysis of C_3S contribute to the difference between C_3A-gypsum pastes and those of Portland cement.

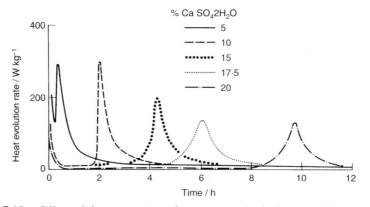

Fig. 7.13.   Effect of the proportion of gypsum on the hydration of C_3A–gypsum pastes (Tenoutasse, 1969) (water/solid ratio = 2)

Mehta (1976) considered that the morphology of crystalline ettringite was not compatible with it forming a layer around C_3A particles giving the degree of retardation observed. A number of alternatives to ettringite have been suggested as responsible for the reduction in the initial hydration rate of C_3A and equilibrium diagrams for the C–A–$\bar{S}$–H system indicate that several metastable phases could be formed according to local conditions, such as sulfate concentration at the solid–aqueous solution interface (Brown, 1989). Microstructural studies, using conventional SEM and TEM to detect and identify hydration products in a paste, are handicapped by extreme sensitivity of hydrous phases to the heating effect of an electron beam in a high vacuum, although relicts of ettringite have been detected by TEM in thin sections (Section 8.1). Scrivener and Pratt (1984) examined C_3A hydration in the presence of gypsum in an environmental cell using HVEM and found that after 10 min apparently amorphous 'filmy material' existed between a C_3A surface and small rods of ettringite about 1 μm from it.

The longer term reaction (up to 20 h) between C_3A and sulfate ions in the presence of calcium hydroxide was found to proceed at a constant rate at high water/solids ratios, by following the disappearance of sulfate ions from solution (Lawrence, 1966). He suggested that this showed that ettringite continuously separated from the C_3A surface, since thickening of a surface layer would cause a progressive decrease in reaction rate. Barnes *et al.* (1996) examined the hydration of a C_3A/gypsum paste in an environmental cell using a high intensity X-ray beam generated in a synchrotron and an energy dispersive method for the rapid recording of diffraction patterns (SR-EDD). They found that ettringite was formed after 50–100 s. No crystalline intermediate was identified but an increase in background intensity in diffraction patterns indicated the presence of an amorphous phase.

The above observations suggest that a two-stage hydration process may obtain: the formation of a metastable product with a composition dependent on conditions at the C_3A surface, followed by rapid dissolution at the high water boundary of the layer and crystallisation through solution to produce ettringite. The rates of these processes and the morphology of the ettringite would be expected to depend on the water/ solids ratio, especially at low values; the concentration and diffusion of sulfate ions; and the calcium ion concentration and pH of the aqueous phase. Eventually the process will be controlled by a reduction in free water content and diffusion involving a progressively thickening layer of ettringite (Brown and LaCroix, 1989).

From the effect of temperature on the zero order kinetics found for the formation of ettringite in a cement paste at a water/cement ratio of 10, Lawrence derived an apparent Arrhenius activation energy of 38 kJ/mol. Other reported values have ranged from 5 to 50 kJ/mol, the lower value implying diffusion as the rate-controlling process. As suggested for C_3S (Section 7.2.1.1), discrepancies are possibly explained by differences in experimental procedure, such as the time interval studied and the water/ solids ratio employed. At the high water/cement ratio used by Lawrence, the influence of a product layer on diffusion of reacting species would be much less than that at normal ratios (0.3–0.5). Lawrence calculated that the formation of ettringite made a contribution of 0.4 W/kg at 25° to the total heat evolution of his samples, indicating that C_3A makes a significant contribution to peak II in Fig. 7.1.

If, as is common in a cement paste, unreacted C_3A remains when the supply of sulfate ions runs out, then there is a surge in its hydration, and ettringite is progressively replaced by a monosulfate (AFm phase) $C_3A.C\bar{S}.H_{12}$. If there is a sufficient excess of C_3A then C_4AH_{13} ($C_3A.CH.H_{12}$) is also produced, in solid solution with the monosulfate.

$$C_6A\bar{S}_3H_{32} + 2C_3A + 4H \longrightarrow 3C_4A\bar{S}H_{12} \tag{7.5}$$

The increased rate of C_3A hydration is identified with peak III in Fig. 7.1. The time at which it appears depends on the ratio of C_3A to gypsum or hemihydrate and also on the reactivity of these phases. Monosulfate can sometimes be detected early in hydration when sulfate ions remain in the aqueous phase, presumably as a result of a localised shortage. Similarly, some ettringite may be found where overall equilibrium would not predict its survival.

7.2.4 Calcium aluminoferrite

The hydration products formed by the ferrite solid solution (xC_2A $(1-x)C_2F$) are usually described as being similar to those formed by C_3A

but with Fe^{3+} partly substituted for Al^{3+}. Reactivity increases with increasing values of x. This phase is deficient in lime compared to its hydration products. If there is a supply of lime from the hydration of C_3S then iron substituted hexagonal hydrates $(C_4(AF)H_{13})$ and cubic hydrogarnets are formed but if not then disproportionation occurs. The lower 'lime content' product appears to be hydrous ferric oxide, originally suggested on the basis of its red-brown colour (Kalousek, 1974). It is sometimes referred to as ferrihydrite, FH_3, although this implies too precise a water content and purity.

From an examination of changes in the composition of the aqueous phase, Brown (1987) concluded that initial hydration of the ferrite phase in the presence of lime and gypsum involved the formation of iron and alumina rich gels at the ferrite surface, low in calcium and almost free of sulfate. Further early reaction occurred by a through-solution mechanism to form an almost iron free ettringite and an iron-rich gel containing some calcium. Subsequent formation of an AFt phase results in the decline in hydration rate and the reaction is considered to become diffusion controlled (Fukuhara *et al.*, 1981).

The effect of changing the proportion of gypsum in a hydrating paste of C_4AF is to produce a set of heat evolution plots equivalent to those in Fig. 7.13 (De Keyser and Tenoutasse, 1968). The second peak, as with C_3A, is the result of a surge in the rate of hydration of the anhydrous compound, resulting in the formation of an AFm phase. There is an increase in the A/F ratio accompanying the formation of the AFt phase, and again in its conversion to an AFm phase, the absolute content of iron in both phases increasing with increasing iron content in the anhydrous ferrite (Tong and Yang, 1994). The difference in the behaviour of iron and aluminium in these reactions is ascribed to the very low solubility (and hence low mobility) of the hydroxide of iron at high pH.

7.3   Hydration of Portland cement

It has often been suggested that to a first approximation the major phases in Portland cement can be considered to hydrate independently. While this may be true in the later stages of hydration when limited amounts of free water remain, in the early stages significant interactions have been found. This is not surprising when it is remembered that most particles larger than about 3–5 μm are polymineralic, that the distribution of the interstitial phases is particularly wide, and that initially at least through-solution processes predominate. An additional factor not present in studies of the pure phases is the extensive and varied levels of solid solution occurring in commercial clinkers. Taylor (1997) has provided a particularly detailed survey of the techniques that have been applied to the investigation of Portland cement hydration, the results obtained and interpretations suggested. In this account, emphasis will be on the processes occurring in the time-scale of the major peaks in reaction rate indicated by conduction calorimetry (Fig. 7.1). The composition and morphology of mature pastes will be considered in Chapter 8.

Several chemical reactions contribute to the initial heat evolution when a cement is mixed to a paste with water, because all the major phases react with it (peak I) and, for the aluminate and ferrite phases, reaction occurs to some extent before the dissolution of calcium sulfate and the formation of calcium hydroxide (from the alite) are sufficient to regulate their hydration. In most modern cements the rapid dissolution of alkali sulfates and, if present, calcium langbeinite (Section 1.4) contribute a significant proportion of the sulfate ion. The nature and rate of these reactions, and hence the degree of hydration resulting at this stage, are sensitive to the rate of solution of the calcium sulfate and to the time and intensity of the mixing process employed, as well as the fineness of the cement.

Uncombined lime and calcium sulfate hemihydrate also react exothermically with water (Table 6.2) and these reactions continue to some degree into the dormant period. If the rate of supply of dissolved calcium and sulfate ions is too low, unretarded hydration of C_3A will occur and flash set is possible. However, if the supply of these ions is too rapid, then the supersaturation with respect to gypsum may rise to a point at which it crystallises, possibly causing false set (Section 6.5). Anything such as aeration which reduces the reactivity of the C_3A may induce false setting tendencies in a cement. Peak I sometimes has a shoulder on it which can be linked to the excessive early stiffening of a paste and prolonged rehydration of calcium sulfate.

On balance, it seems that a major contribution to the heat evolved under peak I comes from the hydration of hemihydrate or if the cement is under-retarded, from the hydration of C_3A. Early reaction soon subsides to a low but detectable level which can be followed by determination of the water bound in ettringite. At this stage a layer of apparently amorphous material, rich in alumina, was observed by Scrivener (1984) in an HVEM examination of a hydrating paste in an environmental cell. Short rods of an AFt phase were observed at its surface and in the aqueous phase (Fig. 7.14). Sujata and Jennings (1992), using ESEM, found that the layer on the particles of a Portland cement was significantly thicker than that formed on the particles in a C_3S paste.

The level of bound water reached during the slow reaction to the end of the dormant period increases with an increase in the levels of C_3A and alkali sulfates in the cement (Bensted and Bye, 1975) and with an increase in the mill temperature during grinding (Bensted, 1982). The dormant period ends with an acceleration in the increase of water bound in C–S–H

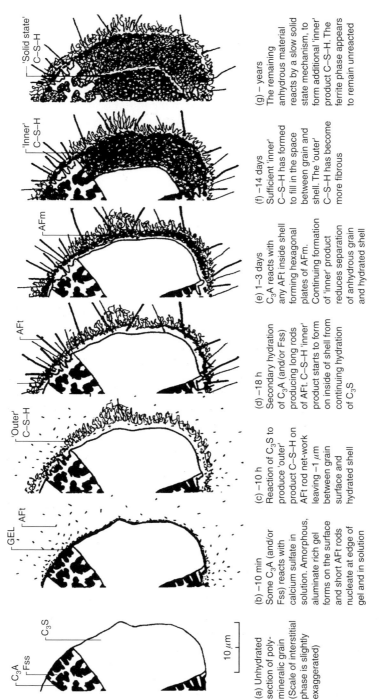

Fig. 7.14. Hydration of a polymineralic grain of Portland cement. Note the persistence of high iron content regions (courtesy of K.L. Scrivener)

(a) Unhydrated section of polymineralic grain (Scale of interstitial phase is slightly exaggerated)

(b) ~10 min Some C_3A (and/or Fss) reacts with calcium sulfate in solution. Amorphous, aluminate rich gel forms on the surface and short AFt rods nucleate at edge of gel and in solution

(c) ~10 h Reaction of C_3S to produce 'outer' product C–S–H on AFt rod net-work leaving ~1 μm between grain surface and hydrated shell

(d) ~18 h Secondary hydration of C_3A (and/or Fss) producing long rods of AFt. C–S–H 'inner' product starts to form on inside of shell from continuing hydration of C_3S

(e) 1–3 days C_3A reacts with any AFt inside shell forming hexagonal plates of AFm. Continuing formation of 'inner' product reduces separation of anhydrous grain and hydrated shell

(f) ~14 days Sufficient 'inner' C–S–H has formed to fill in the space between grain and shell. The 'outer' C–S–H has become more fibrous

(g) – years The remaining anhydrous material reacts by a slow solid state mechanism, to form additional 'inner' product C–S–H. The ferrite phase appears to remain unreacted

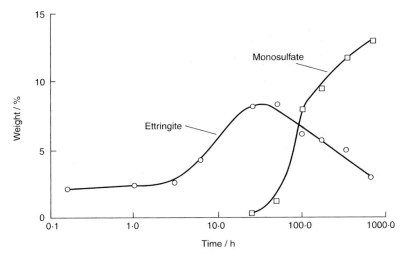

Fig. 7.15. Formation of ettringite and monosulfate in the hydration of a Portland cement followed by quantitative ^{27}Al MAS NMR (Skibsted et al., 1997)

and the AFt phase and a rapid stiffening of the paste. Hydration of C_3S and C_3A then make the major contribution to the heat evolved in Peak II (Fig. 7.1). The consumption of the anhydrous phases can be followed by QXDA, although the relationship with heat evolution makes calorimetry a convenient method of following the progress of the composite reaction.

Skibsted *et al.* (1997) employed ^{27}Al NMR to follow the formation of ettringite and its subsequent conversion to monosulfate in a Portland cement paste hydrated at a water/cement ratio of 0.5 (Fig. 7.15). The first stage involves a change of coordination number of aluminium from 4 to 6. The pattern of two-stage consumption of C_3A was in agreement with that observed by QXDA in a paste of this phase and gypsum reacting in the presence of calcium hydroxide (Kuzel and Pollmann, 1991), suggesting that its reaction in the cement paste was largely free of interference from the additional phases. Premature formation of an AFm phase has been observed at lower water/cement ratios and with finer, 'rapid hardening' (52.5 N) cements (Section 9.2).

The rate of hydration of alite, which makes the major contribution to peak II, is accelerated by the presence of the alkali sulfates which dissolve rapidly. Sodium and potassium ions in solid solution in the clinker phases are released much more slowly. The alkali ions have two effects: they reduce the amount of calcium in solution and raise the pH above the value of approximately 12.5 for saturated calcium hydroxide solution to values above 13, depending on cement composition. Both alkali ions and sulfate ions are retained in C–S–H by adsorption and possibly in the case of the former by incorporation in interlayer sites. Aluminium is widely distributed in the products of hydration of Portland cement and can replace some of the silicon in bridging tetrahedra of the silicate anion in C–S–H.

After 10 h, Scrivener found that the hydration of alite and C_3A in a cement paste had produced a shell of C–S–H on a network of AFt rods around the particles, but, unlike C_3S hydrating alone, a gap of approximately 1 μm existed between the shell and the surface of the cement grains (Fig. 7.14(c)). At normal water/cement ratios, the shells around the individual grains then coalesced to an extent which changed the fracture pattern from intergranular to one involving the breaking of the shells.

In the hydration of the Portland cement sample examined by Skibsted et al., the proportion of ettringite peaked after 12 h and monosulfate formation began. At this stage the concentration of calcium and sulfate ions in a pore fluid falls to very low levels so that continued hydration of the cement takes place in a solution principally of alkali hydroxides. Whether, and at what time, a peak III is observed depends primarily on the A(F)/$\bar{S}$ ratio of the cement. The sample giving the result in Fig. 7.1 contained ca. 13% C_3A (Bogue) and exhibited a marked peak, but many commercial cements (C_3A more commonly 7–10%) produce either no discernible peak or only a shoulder on peak II.

Identification of reactions responsible for minor thermal effects, such as a shoulder on peak II, requires examination of a series of samples spanning the time at which it occurs. Scrivener found that the formation of secondary ettringite was associated with such a shoulder and it differed from that produced initially in forming long needles, while after 1–3 days C–S–H and platelets of AFm had formed inside the hydration shell (Fig. 7.14(d), (e)). The space between shell and anhydrous grain was then progressively filled and subsequently, final slow hydration of alite formed more 'inner product', by an apparently solid state process. Evidence that this mechanism is involved in the hydration of belite has been provided by the observation that its characteristic lamellar structure (Section 4.3) is retained in old, fully hydrated Portland cement pastes (Scrivener, 1986).

The consumption of each of the solid phases in a hydrating cement paste can be followed by QXDA and an example is given in Fig. 7.2. Although the major difference indicated between the curves for C_3S and C_2S is usual, the relationship between those of C_3A and C_3S can change significantly. Rates of hydration depend on the intrinsic reactivity of a phase, the accessibility of it to water in a polymineralic clinker fragment and the effect of a product layer on this and, plotted as fraction remaining, the amount of it present initially. It must be remembered that doubling the C_3A content of a cement does not simply double its rate of consumption.

The microstructure of the products of hydration of the ferrite phase in a cement paste is influenced by the significant quantity of ions in solid solution as well as the limited solubility (and hence mobility) of iron at the pH obtaining. Regions of high iron content thus persist (Fig. 7.14). Rodger and Groves (1989) using TEM found almost circular areas of poorly crystalline material rich in iron, some 200 nm in diameter, intermixed with needle-like crystals and C–S–H. Typical analyses were $C_3A_{0.3}F_{0.3}SiH_x$. The needles appeared to be a hydrotalcite derivative, rich in magnesium from the ferrite solid solution (Section 1.3.4). The mineral hydrotalcite has the composition $[Mg_{1.5}Al_{0.5}(OH)_4](CO_3)_{0.25}H_2O$ and is structurally related to brucite, $Mg(OH)_2$; a wide range of compositions is known and other cations and anions of the right size can substitute for the magnesium, aluminium and carbonate ions.

An apparent Arrhenius activation energy E_A, perhaps better described as a temperature coefficient of hydration rate, can be derived for Portland cement hydration from the results obtained with a conduction calorimeter. Either the Avrami-Erofe'ev equation may be used to determine rate constants from cumulative heat curves obtained at several temperatures (5–50° say), or more simply the height of peak II may be determined as a function of temperature. The second procedure amounts to the substitution of maximum reaction rate for a rate constant but it yields similar results (40–50 kJ/mol for typical commercial cements). Note that in the calculation of E_A the units of the rate constant cancel. A second method of obtaining a substitute for a rate constant is possible because the cumulative heat curve has an almost linear section corresponding to the slow change in heat liberation rate near the top of peak II. From the effect of temperature on the slope of this linear section, Ma et al. (1994) obtained an E_A of 39.0 kJ/mol for a sample of Portland cement. However, in an examination of the effect of temperature on the rate of hydration of a Portland cement, Kjellsen and Detwiler (1992) found that E_A decreased as the degree of hydration increased. It fell from a steady 45 kJ/mol at 20–30% hydration to approximately 10 kJ/mol at 70%, implying a change from a chemical, through-solution process, to diffusion as the rate-controlling step.

7.4   Hydration at elevated temperatures

Most concrete is cured at ambient temperatures, although that at the core of a large structure may experience a temperature rise of 30–40°. In a factory making pre-cast concrete units, curing is often accelerated by using steam at elevated temperatures or even, as in the production of some lightweight concrete blocks, steam under pressure in an autoclave. In such processes hydration reactions are accelerated. Aitken and Taylor (1960) found that C_3A and C_3S in neat cement pastes had completely reacted in 1 and 3 days, respectively, at 100–105° in saturated steam. The degree of silicate polymerisation and crystallinity of C–S–H increase with increasing temperature. At temperatures above about 150°, C–S–H is progressively replaced by $\alpha C_2 SH$ with a loss in compressive strength. Fine sand or fly ash may be incorporated in the mix, to react with the portlandite produced in the hydration of the cement to form additional tobermorite-like C–S–H and maintain strength. Hydrated aluminates and sulfoaluminates are not found in these products and much of the aluminium is in tetrahedral sites in the silicate anion. In low-lime/silica mixes, crystalline silicates such as tobermorite (C_5S_6H), and above 200° xonotlite (C_6S_6H), can be synthesised directly in an autoclave.

7.4.1 Delayed (secondary) ettringite formation

The production of pre-cast concrete, using steam at atmospheric pressure to accelerate curing, has revealed a problem where items such as railway sleepers (ties) and bridge beams have subsequently been exposed at ambient temperatures to wet conditions. Ettringite is not formed in concrete at temperatures above about 70° and any formed by hydration at initially lower temperatures is slowly decomposed (Kalousek and Adams, 1951). Using SR-EDD, Barnes et al. (1996) examined the rate of decomposition of ettringite leading to the loss of sulfate with the formation of an AFm phase in the range 70–90°. This was then transformed into a calcium aluminate hydrate after prolonged exposure (around 5 h) to the highest temperatures. Both of these transformations were shown to be reversible.

After an induction period, which may be as long as several years, slow expansion of a steamed cured concrete stored in water at ambient temperatures may be observed and well formed crystals of ettringite in bands 10–50 μm thick develop around aggregate and sand particles. At the high temperature of cure, some aluminium becomes bound as hydrogarnet and some replaces Si in C–S–H, but apparently much remains available for recombination at low temperatures with sulfate ions remaining in the pore water or adsorbed in the C–S–H. Microanalysis has demonstrated the diffusion of initially dispersed sulfate and aluminate species in the development of this ettringite (Yang et al., 1996). It can sometimes be detected in voids and cracks after a few days.

Serious, deep cracking in a concrete results if expansion approaches about 1%. Since concrete deterioration was associated with delayed ettringite formation (DEF) by Ludwig and co-workers in the 1980s, laboratory investigations have sought to establish under what conditions of cure and use, and with which materials, a concrete will be susceptible to the phenomenon. The variables which have been shown to influence induction time, the rate and degree of expansion and whether serious cracking follows include: cement composition and fineness; aggregate type; curing regime (rate of heating, temperature reached and time at maximum); and conditions of subsequent exposure of a mortar or concrete. In test samples, expansions observed are much lower (or much delayed) if a limestone aggregate is used or if a concrete is stored in water-saturated air rather than under water. In field studies of concrete it is not always easy to separate DEF from cracking resulting from thermal, freeze-thaw or humidity cycles (Stark and Bollmann, 1997).

There is no general agreement on the relative significance of variables in the chemical composition of commercial Portland cements. Factors cited as positively correlating with expansion have included C_3A, C_3S, SO_3 and alkali contents, as well as cement fineness. Kelham (1997) found a pessimum SO_3 content of ca. 4% for his samples. Because of the effect of alkali concentration in pore water on ettringite stability at high temperatures (Brown and Bothe, 1993) and its subsequent effect on ettringite crystallisation at ambient temperatures, expansion has also been linked with the reaction of alkali with siliceous aggregate (Section 9.2).

One hypothesis is that expansion results from the crystal growth pressure exerted by the observed ettringite crystals. The induction period is said to be the result of the slow increase in supersaturation of the pore solution with respect to this phase. At a sufficiently high level, favoured nucleation sites are provided by preformed cracks and the voids around aggregate particles created by the differential thermal expansion of the cement paste and aggregate. Fu *et al.* (1994) stressed the importance of ettringite formation at crack tips where very high stresses could be generated and Diamond (1996) observed that growing ettringite crystals could penetrate mica lamellae in steam cured concrete.

An alternative suggestion is that expansion occurs as a result of the formation of ettringite finely dispersed throughout the volume of the paste fraction of a concrete and that this expansion accounts for the formation of the rather uniform voids around the aggregate particles. The observed, larger crystals of ettringite are considered to result from secondary recrystallisation (Ostwald ripening). Lewis and Scrivener (1997) obtained evidence supporting this mechanism. They found no relationship between the rates of expansion and amount of ettringite found, and this phase was formed in some mortars which exhibited no expansion. A significant finding in this study was an apparent link between the microanalysis of the paste fraction of their mortars and subsequent expansion (Fig. 7.16), although they emphasised the need for more work before this could be regarded as diagnostic.

Whatever the finally agreed mechanism, the microstructure of the paste after heat treatment and the absorption of water into denser regions of the gel (to provide the high level bound in ettringite) are factors potentially

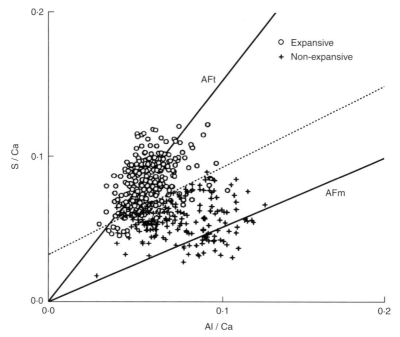

Fig. 7.16. Plot of atomic ratios, S/Ca against Al/Ca, determined by X-ray microanalysis in hydrated material in mortars 1 day after heat treatment: the broken line separates mortars that subsequently expanded in water from those that did not (Scrivener and Lewis, 1997)

influencing expansion. As Taylor (1994) pointed out, expansion accompanying a chemical reaction will probably involve changes in porosity as well as differences between the specific volumes of reactants and products. Detailed accounts of DEF have been given by Lawrence (1995) and Diamond (1996). The latter drew attention to the parallelism with long-term expansion of concrete which has not been steam cured, and pointed to the fact that in the past 50 years total SO_3 contents of commercial cements have risen from around 2% to as much as 4%. This increase is often linked with an increase in the level associated with alkali in clinker produced in modern, thermally efficient plant.

7.5 Concluding remarks
Recent research has yielded considerable insight into the mechanisms of the hydration of the individual phases in Portland cement. Two factors emerge as pre-eminent in determining initial hydration rates:

  (*a*) the intrinsic reactivity of the solid phase (polymorph) — determined in part by the conditions it has experienced in its formation such as cooling rate after firing, as well as by solid solution and other lattice defects;

(b) the nature of the aqueous phase — the concentration and availability of a supply of calcium and hydroxyl ions and also sulfate ions in the case of the aluminate and ferrite phases.

Initial through-solution hydration results in the rapid precipitation of insoluble material close to the anhydrous grains. These products are amorphous but subsequently either crystallise or develop morphologies which make acceleration of further hydration possible. Eventually the increasing amount of the hydration product and the decreasing content of free water result in diffusion, rather than one or more of the 'chemical' steps indicated in Fig. 7.12, becoming rate controlling. This is reflected in a fall in the apparent Arrhenius activation energy. The time-scale of these changes is dependent on both the particle size distribution of the cement and the water/cement ratio chosen, and these must be taken into account when seeking correlations between degree of hydration, curing time and the physical properties of hardened pastes.

There is evidence that hydration does not occur uniformly as implied in Fig. 7.12, but preferentially at sites where a localised lattice strain exists, such as dislocations and twin planes (Section 1.3). In Portland cement pastes, the existence and accessibility of the individual phases in polymineralic grains further complicates the hydration process. The initial amorphous layer resulting in a dormant period is rich in alumina, calcium and silica. Not surprisingly, subsequent hydration leads to the formation of product layers of variable thickness around many grains, as well as different thicknesses on different grains (Section 8.1).

In spite of the complexity of the hydration reactions at the single particle level, averaging in a large assembly of particles means that relatively straightforward kinetic equations apply to degree-of-hydration/ time $(\alpha - t)$ plots. Properties which provide a measure of α, such as cumulative heat and chemical shrinkage, can replace it. The equations of Avrami-Erofe'ev (7.2) and those proposed by Knudsen (1984), which link α and t, have been employed in hydration modelling procedures for predictive purposes. The former was originally proposed for solid state reactions involving acceleratory and deceleratory processes but can fit others with a sigmoidal $\alpha - t$ relationship. Constants for use with Avrami-Erofe'ev equations for the four main phases of cement were determined empirically by Taylor (1987).

Knudsen (1984), on the other hand, developed equations for long-term hydration (beyond the acceleratory period, $\alpha > 15\%$) taking into account the dispersion of particle sizes in a typical cement. Observed chemical shrinkage measurements indicated two limiting cases; in one, the hydration ratio $(\alpha/1 - \alpha)$ had a linear dependence on time $(\propto t)$, in the other, a parabolic one $(\propto t^{\frac{1}{2}})$. His results suggested that parabolic kinetics were favoured by a high level of available Ca^{2+} ions, from gypsum or added calcium chloride, at the commencement of hydration. The existence of linear kinetics, which Knudsen observed in the hydration of a white cement clinker, suggests that the build-up of a layer of hydration products did not limit access of water to hydrating surfaces in this system.

8. The nature of hardened cement paste

The idea that the strength developed in a hardened Portland cement paste is not determined by the degree of hydration of the cement alone was introduced in Section 7.1. The dependence of strength on initial water/cement ratio is an obvious manifestation of this but more subtle strength–constitution relationships are also involved. In this chapter, some physical characteristics of hardened cement paste and their influence on engineering properties, such as compressive strength and drying shrinkage, will be considered.

8.1 Microstructure of hardened cement paste (hcp)

The hydration products of Portland cement range in type from coarsely crystalline portlandite (CH) to almost amorphous calcium silicate hydrate (C–S–H). Characterisation of the resulting composite involves the determination of the texture of the individual phases present and also of the overall assembly of phases. The term 'texture' encompasses the perfection, size and form of individual crystals as well as the way they are packed together and the nature of the porosity (shape and volume–size distribution) this produces. Using the results of microanalyses of clinker and hydration products, Taylor (1997) set out the basic assumptions needed in a calculation of the phase composition of a hardened cement paste. For a saturated paste obtained with a typical Portland cement hydrated for 14 months at a water/cement ratio of 0.5, he calculated *volume percentages* for: C–S–H 48.7%; CH 13.9%; AFm 11.1%; AFt 3.6%; pores 16.0%; unhydrated clinker 2.6%; the remainder consisting of minor hydration products and some carbonated material.

With the exception of calcium hydroxide, even the crystalline phases identified by X-ray diffraction are best examined in the electron microscope because they possess at least one very small dimension.

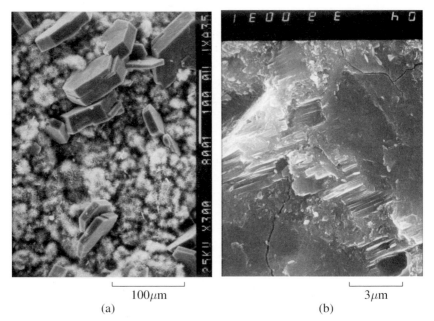

100μm	3μm
(a)	(b)

Fig. 8.1. Scanning electron micrographs of fracture surfaces in hardened cement pastes: (a) three-day old paste containing well-defined portlandite crystals; (b) fracture through dense groundmass in an eight-month old paste

The earliest studies, using TEM, involved crushing a hardened paste and examining the dispersed fragments. Although this technique revealed the form of individual constituents, it gave no indication of their packing in the paste. In the 1970s, the topographical images produced in the SEM greatly increased appreciation of the overall texture of hardened pastes of both cement and the individual silicates, especially since they could be examined over a wide range of magnification. In the 1980s, the development of BSE imaging coupled with X-ray microanalysis provided the basis for a method of quantitative characterisation of both phase and chemical element distribution, as well as porosity, in hardened pastes.

In the early stages in the hardening of a paste, hexagonal prisms of ettringite and fibrous C–S–H form around the clinker grains and regularly shaped calcium hydroxide crystals are found between them (Fig. 8.1(a)). In the later stages of hydration considerable intermixing of the different phases occurs producing a relatively featureless material, usually referred to as *groundmass* (Fig. 8.1(b)). Distinct grains, the 'phenograins' of Diamond and Bonen (1995), remain dispersed in the groundmass and they consist of either portlandite or polymineralic clinker particles hydrated in situ to a wide range of degrees, with individual hydration shells frequently variable in thickness. In concrete, massive formations of CH develop, many associated with the interface (*transition zone*) between paste and aggregate, steel reinforcement, or pretensioning wires. In pastes hardened at very low water/cement ratios, diffusion and crystal growth of CH can be expected to be limited and Groves (1981) found clusters of very thin CH platelets embedded in C–S–H in hcp produced at a ratio of 0.2.

Pores with diameters down to 0.5–1 μm (many filled with stabilising resin) are resolved in a BSE image and different levels of greyness permit the identification of partially hydrated clinker particles and portlandite crystals. Coupling SEM with an image analyser thus makes it possible to determine area fractions for: unhydrated and hydrated clinker, CH crystals (many of which are coarse aggregates of platelets) and pores. The application of automated image analysis to the examination of hcp microstructure is described by Darwin and Abou-Zeid (1995). Scrivener *et al.* (1987) found that the anhydrous clinker fraction derived from BSE images correlated well with that calculated from loss on ignition determinations, but irregular distribution of CH particles in their samples necessitated time-consuming counting of large areas to obtain a satisfactory result. Wang and Diamond (1995) examined pastes hydrated for 100 days in CH solution at water/cement ratios of 0.3 and 0.45. The magnitude of the difference in the area fraction of unhydrated clinker remaining was significant, 6.0% at the higher ratio but 17.4% at the lower.

Diamond (1972) found that in X-ray microanalysis, portlandite was distinguished from C–S–H in the groundmass by the fact that only the latter contained significant amounts of sulfur (as sulfate). Finely crystalline phases can only be picked out by SEM where they protrude into pores and then they are intermingled with fibrous C–S–H. AFt phases are found as narrow hexagonal prisms or rods up to 5 μm long; AFm phases as platelets less than 0.1 μm thick. Since fracture surfaces examined in SEM tend to follow a concentration of pores, the number of these forms seen is unlikely to be typical of their overall abundance. In the X-ray microanalysis of samples of a paste, it is probable that more than one of these finely divided phases will lie in the volume irradiated by the electron beam. In such cases, determination of two atomic ratios, silicon/calcium and aluminium/calcium for example, and plots of one against the other are aids to interpretation (Fig. 8.2).

Early application of SEM revealed a significant number of holes in the microstructure of hcp samples, apparently where particles of clinker had completely dissolved (Hadley, 1972). A later view was that these *Hadley grains* were artefacts formed by particle pull-out in specimen preparation, but recently further examination of this phenomenon has suggested that it is a result of the through-solution mechanism in early hydration (Kjellsen *et al.*, 1996).

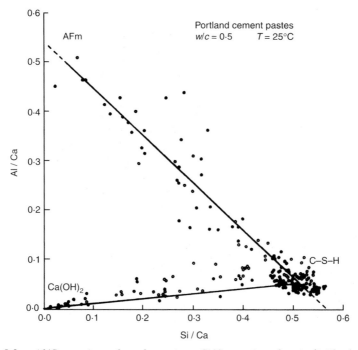

Fig. 8.2. Al/Ca ratios plotted against Si/Ca ratios for individual X-ray microanalyses of typical hardened cement pastes: results grouped around the joins linking specific compositions are produced by simultaneous excitation of more than one phase by the electron beam (courtesy of H.F.W. Taylor)

Four forms of C–S–H were described by Diamond (1976) but the principal distinction is that between the relatively porous, apparently fibrillar, outer hydration product (OP) filling the space between grains of clinker and CH and that formed as 0.1–0.3 μm equant grains of inner product (IP) occupying space within the boundaries of the original, larger clinker grains. In hydration this differentiation is usually established within 24 h. Jennings and Pratt (1980) found that in a C_3S paste thin foils of C–S–H rolled up on drying, except where those from adjacent clinker particles overlapped, and it has been suggested that the form of C–S–H particles in hcp is determined by the space available as they develop and subsequently dry. Scrivener distinguishes between that C–S–H deposited up to 14 days in the gap created by early through-solution hydration and that formed much more slowly by 'solid state' reaction (Fig. 7.14). This final in situ hydration must involve removal (leaching) of calcium and silicate ions from the clinker.

The greater resolution of TEM and the use of ion beam thinned sections of hcp reveal the microstructure of IP (Fig. 8.3). Richardson and Groves

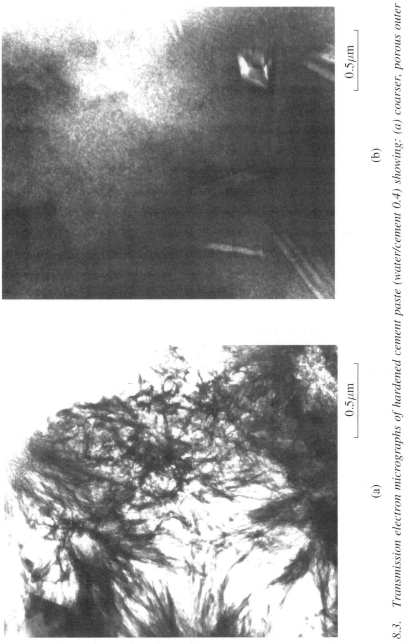

(a)

(b)

Fig. 8.3.   Transmission electron micrographs of hardened cement paste (water/cement 0.4) showing: (a) coarser, porous outer C–S–H after 1 week; (b) the finer grained inner product after 1 year (courtesy of I.G. Richardson)

(1993) demonstrated by electron diffraction the absence of crystalline phases in the areas of paste they examined using TEM and X-ray microanalysis. The least thermally stable phase (AFt) was present as relicts of the original acicular crystals, which had the correct atom ratios for the elements Al, Fe, Ca and S. A residual amount of the AFt phase remained as the AFm phase replaced it in samples hydrated for longer periods. Outer product C–S–H was intermixed with large amounts of CH and lesser amounts of AFm platelets. These phases were occasionally found in IP C–S–H, together with some magnesium- and iron-rich phases, presumably as a result of enrichment of these two elements in residues after selective leaching. Ca/Si ratios were variable over a similar range in OP and IP C–S–H. No systematic variation with the age of a paste was observed but they found evidence of a bimodal distribution of values at early ages, as had Rayment and Lachowski (1984). In the several one-year old samples which Richardson and Groves examined, Ca/Si ratios ranged from 1.61 to 1.90.

Resolution of the structure of C–S–H at the finest level has been achieved using field emission TEM to produce lattice images. Viehland *et al.* (1996) noted that C–S–H pastes gave diffuse electron diffraction rings, indicating the presence of short range order, and that the diffuseness varied from one area of a sample to another. Lattice images revealed nanocrystalline regions (ordered atomic patterns up to 5 nm) and regions of short range order (up to 1 nm) in an amorphous matrix. Nanocrystallites (>10 nm) of CH could be distinguished from C–S–H. Not surprisingly, the volume fraction of nanocrystalline regions was significantly greater in samples cured at 80° than in those cured at room temperature.

8.2   Surface area and porosity of hardened cement paste

Powers, Brunauer and their collaborators in the 1950s recognised that the physical properties of hardened cement paste are dominated by the high surface area and surface energy of C–S–H, since it exists as an assembly of very small, almost amorphous particles. They used the term tobermorite gel to describe C–S–H (Section 7.2.1). The solids content of hcp is, however, much higher than it is in what is normally described as a gel (of bentonite, for example), but the constituent particles have the right order of fineness. For a specific surface of $200 \, m^2/g$, a particle with dimensions $2 \, \mu m \times 0.5 \, \mu m$ would have a thickness of only about $0.005 \, \mu m$, assuming a density of $2500 \, kg/m^3$. Because of its high solids proportion, C–S–H is sometimes described as a xerogel, the product of removing the dispersion medium from a gel.

8.2.1   Surface area

The specific surface area of hcp is usually determined by measuring the adsorption of either water at 293K, or nitrogen at 77K, and using the BET equation to calculate the surface area from the results. Two complications arise:

(a) the outgassing procedure preceding adsorption measurement can only be arbitrarily defined as there is no clear distinction between the removal of adsorbed and chemically bound water;

(b) the value obtained for the surface area depends on the adsorbate used.

D-drying (Section 7.2.1) is frequently used, although it has been claimed that other more rapid methods produce a similar dried and outgassed state. Nitrogen is normally the preferred adsorbate for hydrous oxides because its molecular packing is less sensitive than that of water to the detailed chemical characteristics of the surface. Brunauer (1970) found that for samples of D-dried hcp, nitrogen yielded surface areas which decreased from around 150 to $50\,m^2/g$ as the water/cement ratio employed decreased from 0.7 to 0.35. In contrast, water adsorption gave surface areas which were all close to $200\,m^2/g$. Determinations of the specific surface of hcp by low angle X-ray scattering (Diamond and Winslow, 1974) supported the uniformly high values given by water adsorption.

The much higher values for surface area obtained by water vapour adsorption at lower water/cement ratios suggest that there are pores in C–S–H which are accessible to water but not to nitrogen. Three explanations of this have been suggested:

(i) the water molecule (0.3 nm) is smaller than the nitrogen molecule (0.42 nm);

(ii) diffusion rates for water molecules at 293K are much higher than they are for nitrogen at 77K, so that if many of the pores in hcp possess very narrow entrances (*ink bottle pores*) then their surface will not be determined by nitrogen, a situation observed with some porous coals;

(iii) the high polarity of the water molecule enables it to separate solid surfaces, including possibly a process of interlayer hydration analogous to that found with swelling clays.

Some resolution of the problem of the different behaviour of water and nitrogen in their adsorption on hcp came with the observation that the low surface area values given by the latter were the result of the sensitivity of hcp to shrinkage when being dried. The effect of this shrinkage, which results from the high surface tension of water in fine pores, is to increase the area of surface to surface adhesion, particularly as drying proceeds from 70% down to 40% rh (Parrott et al., 1980). While the effect can be reversed by re-wetting, the spreading pressure in an adsorbed nitrogen film is insufficient to overcome the adhesion in the area lost. Such shrinkage can be prevented by first using a polar, low surface tension liquid miscible with water to displace it from hcp, followed by a more volatile liquid to displace the first. Litvan (1976) used methanol and pentane, the latter removed by vacuum pumping. Possible interaction of methanol with hcp is prevented by avoiding prolonged exposure but other workers have preferred propan 2-ol.

Rarick *et al.* (1995) discussed the problems of determining and specifying the surface area of hcp. While the obvious reference for the specific surface area of a sample is its weight and composition after outgassing, they pointed out that published values have often been related to the ignited weight and hydrated compositions not given. A third method encountered relates values to the calculated weight of C–S–H present in a sample. The three methods of expressing the result yielded surface areas for water vapour adsorption on a D-dried sample of approximately 195, 238 and 358 m^2/g, respectively.

Jennings and Tennis (1994) noted that, for D-dried pastes, as the surface area determined by nitrogen fell with decreasing water/cement ratio, pore volumes inaccessible to this adsorbate had been found to increase. They developed a model in which cement hydration resulted in the formation of two forms of C–S–H: one with pores accessible to nitrogen and a second with pores only accessible to water. In their model, as water/cement ratio increased, the proportion of the latter decreased. A relationship of the two forms to the outer and inner C–S–H observed in electron microscopy was one suggestion they made and neutron scattering has since provided support for the existence of two forms of this hydration product (Thomas *et al.*, 1998).

Feldman and Sereda (1968) proposed a model for the structure and drying–re-wetting behaviour of C–S–H in which interlayer water and its movement with changes in relative humidity play a significant part, especially below 11% rh. Their model is illustrated in Fig. 8.14 and it will be considered in connection with drying shrinkage in Section 8.3.4. The effect of reversible and irreversible adsorption of water has been discussed in some detail by Beaudoin and Brown (1993). There is certainly a parallel between the difference in the surface areas obtained for hcp using nitrogen and water and the difference for the same two adsorbates with swelling clays. Unfortunately, linking water adsorption with an increase in lattice spacing, as Van Olphen (1965) did for sodium vermiculite, is not possible for the C–S–H in hcp because no suitable lattice spacing is available. However, for a semi-crystalline form of tobermorite (C–S–H I) a relationship between relative humidity and lattice spacing has been observed (Bayliss, 1975; Gutteridge and Parrott, 1977).

8.2.2   Porosity

The above considerations make it clear that for a paste to have a specific surface as high as $200 \, \mathrm{m^2/g}$, its denser parts must contain pores not observed in SEM (width $<$ ca. 0.5–$0.8 \, \mu\mathrm{m}$). Powers and Brownyard (1948) distinguished between two sorts of porosity in hcp: *gel porosity* which is contained within an assembly of C–S–H particles, and *capillary porosity* between the separate regions filled with gel and the crystalline phases. The latter is taken to contain pores of greater width, to be mainly derived from space originally occupied by air or by water not used in hydration, and therefore to depend on the original water/cement ratio. However, within the capillary pore system there must be narrow regions where solid surfaces approach points of contact, so that the pore size distributions for the two systems will overlap. Rarick *et al.* gave a summary of a more detailed classification of pores, linking the role of water in them with the physical properties of a paste which are considered in subsequent sections (Table 8.1).

Powers (1964) defined the total porosity of a paste as that occupied at saturation by evaporable water which he took to be the water removed by D-drying. He assumed that this water has a specific volume V_e of approximately $1 \, \mathrm{cm^3/g}$, although he recognised that it includes some chemically bound water. He then related total porosity (ϵ) to the fraction of cement hydrated (m) and total mass of water (including any originally chemically combined in the cement) in the specimen (w_t):

$$\epsilon = w_e V_e / V \tag{8.1}$$

Table 8.1.   Classification of pores in hardened cement pastes (after Rarick et al., 1995)

Designation	Diameter	Description	Role of water	Properties influenced
Capillary pores	10–$0.05 \, \mu\mathrm{m}$	Large capillaries	Bulk water	Strength, permeability
	50–$10 \, \mathrm{nm}$	Medium capillaries	Menisci generate moderate surface tension forces	Strength, permeability, shrinkage at high humidities
Gel pores	10–$2.5 \, \mathrm{nm}$	Small (gel) capillaries	Menisci generate strong surface tension forces	Shrinkage to 50% rh
	2.5–$0.5 \, \mathrm{nm}$	Micropores	Strongly adsorbed water, no menisci	Shrinkage, creep
	$< 0.5 \, \mathrm{nm}$	Micropores	Structural water involved in bonding	Shrinkage, creep

and

$$\epsilon \sim w_e/V = (w_t - m w_n^o)/V \tag{8.2}$$

where w_n^o is the mass of the non-evaporable (bound) water (w_n) for the completely hydrated state, V the total volume of the paste and w_e the mass of evaporable water. If c is the mass of cement, then multiplying Equation (8.2) by V/c gives an expression containing water/cement ratios:

$$w_e/c = w_t/c - m w_n^o/c \tag{8.3}$$

Using results for well-cured pastes which had a range of initial net water/cement ratios w_o/c, and which were made with a cement for which the water/cement ratio for complete hydration (w_n^o/c) was expected to be 0.227, Powers plotted w_e/c against w_t/c and the results lay on two straight lines with a distinct difference in slope (Fig. 8.4). The line AB has a slope of unity corresponding to a degree of hydration, m in Equation (8.2) of 1 (100%). In these pastes porosity varied only with w_t/c which was related to w_o/c as shown in Fig. 8.4.

The pastes giving results lying on the line OA had been cured under water for up to four years but they were neither hydrated completely nor to the same extent, even when they contained water in excess of that needed chemically. For the cement used to obtain the results in Fig. 8.4 the critical net water to cement ratio was 0.38. This value was apparently necessary to provide sufficient space for the development of the hydration products and Powers concluded that $1\,\text{cm}^3$ of cement requires about

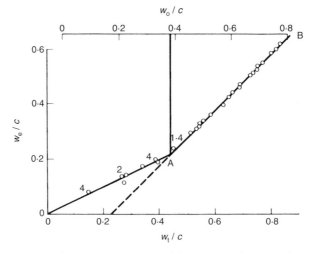

Fig. 8.4. *Relationship between evaporable (w_e), total (w_t) and net (w_o) water contents per unit mass of cement (c) in mature cement pastes (Powers, 1964). Numerals indicate period of curing in years*

$2.2\,cm^3$ of space of complete hydration, that is $1.2\,cm^3$ must be provided by the added water.

Such results indicate that the strength of the gel is greater than the crystallisation pressure of the hydrates since otherwise the paste would be disrupted. The limiting case in Fig. 8.4 corresponds not to zero porosity but to zero capillary porosity, a gel porosity of 0.28 remaining. Mills (1966) also obtained results supporting the idea that lack of available space may stop hydration.

Powers recognised the limitation of determining total porosity by measuring the weight loss in D-drying a paste and subsequently Harris *et al.* (1974), using neutron scattering, confirmed that the drying weight loss included a significant amount of combined water. A problem of failure to differentiate between free and bound water also arises if the solid volume in a sample of dried paste is determined by immersion in water. Calculation of the porosity also requires a value of the apparent, overall, solid volume of the sample, which can be determined by immersion in mercury. To avoid the rehydration and swelling occurring when water is used, methanol or propan 2-ol can be employed as pyknometric fluids.

Water has been shown to yield similar porosities to methanol, if paste samples are equilibrated at 11% rh or above and Sereda *et al.* considered that at this rh most of the interlayer water (X in Fig. 8.14) remains in the paste. Similar results are given by helium pyknometry which is carried out at 2 atmospheres. The determination is completed in 2 min and the slower flow of helium into vacant interlayer sites (if dehydrated) can then be followed. This technique has been used to examine changes in the porosity of pastes as drying is carried out below 11% rh (Feldman, 1972).

8.2.3   Pore size distribution

The size distribution of the accessible pores in hcp may be examined by mercury porosimetry in which the volume of mercury penetrating a sample of paste is measured as a function of increasing pressure (p), which is related to the diameter (d) of the pores just penetrated at that pressure by the equation:

$$p = 4\gamma \cos \theta / d \qquad\qquad (8.4)$$

where γ is the surface tension of mercury and θ its advancing contact angle with the solid.

The limitations of this technique include the usual assumptions that the pores are cylindrical and that for want of information θ is 135°, and also the uncertainty concerning the effect of pressures (commonly up to $350\,N/mm^2$) on the solid. Interpretation of the results is also complicated by the presence of pores possessing narrow entrances since the volume

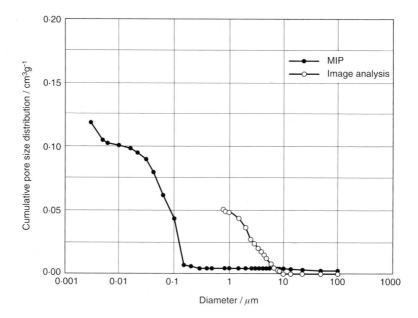

Fig. 8.5. Comparison of pore size distributions in hcp (28 days at water/cement = 0.40) determined by mercury intrusion porosimetry and BSE image analysis (Diamond and Leeman, 1995)

penetrated is recorded as belonging to pores with a diameter equal to that of the neck of an 'ink bottle pore', although a tortuous pore with one or more narrow constrictions along its length would be a more realistic model. Nevertheless, availability and ease of use have meant that mercury intrusion porosimetry (MIP) has been widely employed to demonstrate the influence of water/cement ratio and curing on pore size distribution. MIP does reproducibly give results indicating the anticipated trends and is considered to give a meaningful total pore volume. Beaudoin (1979), using pressures up to $400\,N/mm^2$, found that the volumes intruded by mercury were similar to those indicated by helium pyknometry for pastes hydrated with a water/cement ratio of 0.4 or above. The development of automated image analysis coupled with BSE imaging in SEM has quantified the ink bottle effect somewhat for pores down to 0.5–0.8 μm. Diamond and Leeman (1995) compared pore size distributions determined by BSE imaging and MIP and found that pores between 10 μm and the BSE lower limit were largely undetected by MIP (Fig. 8.5).

Determination of the size distribution of pores with diameters ranging from just above molecular dimensions up to about 40 nm is possible using the data from isotherms from the physical adsorption of a gas. Nitrogen and water vapour are normally employed with cement paste and the

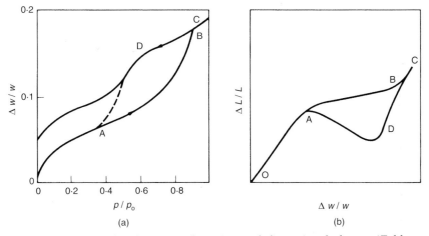

Fig. 8.6. Relationship between adsorption and dimensional change (Feldman and Sereda, 1968): (a) typical water adsorption isotherm for hcp, amount adsorbed per unit weight of sample ($\Delta w/w$) as a function of relative humidity (p/p_o); (b) relationship between adsorption and fractional length change of sample ($\Delta L/L$) (see Section 8.3.4)

method involves measurement of the amount of adsorbate condensed as liquid (capillary condensation) in the pore system of the paste as a function of its relative pressure. The basis of the method and some problems of interpretation are discussed in the Appendix.

A typical water adsorption isotherm for hcp is shown schematically in Fig. 8.6 and, like most published water (but unlike nitrogen) isotherms, it shows a large hysteresis loop (solid line) which does not close until the pressure is reduced to very low values. However, Brunauer and his co-workers (1972) found that if pastes were equilibrated for four to five months at each relative pressure selected for an adsorption measurement, then the hysteresis loop closed as indicated by the dashed line in Fig. 8.6.

Figure 8.7 contains pore size distributions obtained for cement pastes from water and nitrogen adsorption isotherms. A simpler representation of the volumes of fine, medium and coarse pores was used by Parrott and his co-workers (1981, 1988). They defined three ranges of pore size: gel porosity (< 4 nm width); small capillary porosity (4–37 nm width) and large porosity (> 37 nm) and determined the volumes of pores in these ranges by comparing the simultaneous adsorption of methanol on samples of hcp and two samples of microporous glass, with pore widths of 4 and 37 nm. Total pore volumes were determined from the methanol-saturated weights of the specimens. The method assumes that adsorption in the pore systems of both glass and hcp involves the same molecular packing. Plots of the normalised volume fractions of solids and the chosen pore size ranges reveal the

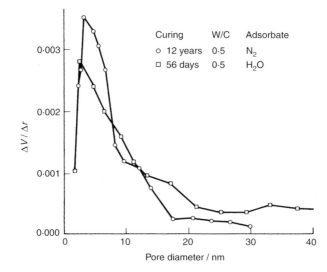

Fig. 8.7. Pore-size distributions of hcp derived from nitrogen and water isotherms (Diamond, 1971). A plot of the derivative is usual for this technique. ΔV is the volume of pores in small chosen intervals in pore radius (Δr)

progressive decrease in the content of larger pores as the fine porosity associated with hydration products replaces it (Fig. 9.5).

It is important to note that the pore size distribution and hence strength and permeability of concrete are influenced by two special effects not present in hcp. These are:

(*a*) the relatively high porosity *transition zones* which develop as coarse crystals of portlandite grow around aggregate particles and reinforcing rods — they are considered to be the result of crystal growth involving diffusion of CH into water which has bled from the adjacent paste during shear and placing of the mix;

(*b*) the presence of a low porosity skin of hcp as a result of compaction and levelling.

The transition zone in concrete has been the subject of much recent interest (RILEM, 1996) and computer modelling of it (Section 8.5) has been described by Bentz *et al.* (1995).

8.3   Physical properties of hardened cement paste
8.3.1   Permeability

The poor tensile and flexural strengths of cement paste make it necessary to improve these properties in concrete by incorporating steel rods, and the ability of the concrete to protect them from corrosion in a wet environment is largely determined by its permeability. This is therefore an important property of structural reinforced concrete although permeability is not incorporated in standards. A high pH is produced in hcp (typically 13.5) and aided by the fact that portlandite tends to crystallise around the steel, this inhibits metallic corrosion. The durability of the concrete depends on how long the passivation of the iron lasts. The passivation is reduced by carbonation of portlandite by atmospheric CO_2, which reduces the high pH inducing it, and by penetration of chloride ions into the concrete, from deicing salts, for example. The rates of these processes increase with increasing permeability of the concrete, which is dependent on its pore diameters and their continuity (Brown et al., 1991).

Only the permeability of hcp to water is considered here. Problems in measuring it include lime leaching, continued hydration of clinker remnants and trapping of air in the pores (Lawrence, 1986). Water rapidly penetrates dry (unsaturated) cement paste but, once saturated, flow through a specimen becomes steady at a low rate ($F\,\mathrm{m}^3/s$) given by a form of Darcy's equation:

$$F/A = Kh/L \tag{8.5}$$

where A is the cross-sectional area, h the drop in hydraulic head (m), L the thickness of the specimen and K the permeability coefficient in m/s.

Powers et al. (1959) pointed out that for a paste containing cement of normal fineness, the continuity of the capillary porosity, which provides the most significant diffusion path, can be eliminated by curing to block it with cement gel if the initial water/cement ratio is less than 0.7. At this ratio, capillary cavities remain but they are separated by gel which contains much finer pores. They concluded that only for water/cement ratios less than about 0.35 could capillary porosity be 'cured out'. At a value of 0.7 they found that a curing time, at normal temperatures for ordinary Portland cement, of about one year was required to block capillary porosity, while at 0.4 only three days were necessary. In the process the permeability coefficient fell by 4–5 orders of magnitude to a value of the order of 10^{-13} m/s. Garboczi and Benz (1991) developed a three-dimensional computer model of capillary pore connectivity (continuity) and found that the capillary porosity at which discontinuity first occurred lay in the range 18–20%, comparable with that found experimentally by Powers et al., 19–26%.

Nyame and Illston (1980) defined a maximum continuous pore radius in the pore size distribution determined by mercury porosimetry, as the radius at which the increase in volume (v) penetrated with unit increase in applied pressure (p) is a maximum. For a range of water/cement ratios from 0.23 to 1.0 and curing times of up to 20 months, they found a linear relationship between the log of the permeability coefficient and the log of the maximum continuous radius. However, the use of MIP to define the latter means that the relationship is unlikely to be absolute. They found it necessary to make a large number of replicate permeability measurements because the coefficient of variation was of the order of 50%. Results for the fall in the permeability of hcp with decreasing water/cement ratio and increasing curing were similar to those of Powers *et al.* The latter considered K within C–S–H gel to be of the order of 10^{-15} m/s. Movement of water within this fine pore system will be by surface diffusion rather than the normal liquid flow possible in capillary pores.

8.3.2 Compressive strength

For practical purposes, the compressive strengths developed by cements in performance tests are presented as functions of curing time and water/cement ratio (Fig. 6.2), and these variables are the principal influences on the porosity of hardened cement paste. A number of equations relating porosity to strength have been applied to data for hcp and good correlations found, at least over some ranges of porosity.

Powers (1958) considered that strength is related to the concentration of solid hydration products in the paste and expressed this in a *gel space ratio* (X), defined by Equation (8.6) in which the gel pores are included in the gel volume:

$$X = \frac{\text{gel volume}}{\text{gel volume} + \text{capillary volume} + \text{entrained air volume}} \quad (8.6)$$

He related strength σ to X by the equation:

$$\sigma = AX^n \quad (8.7)$$

where $n = 2.6$–3 for the range of cement samples examined.

Powers found a linear relationship between log compressive strength and log of the gel space ratio which extrapolated to a maximum strength A at $X = 1$, corresponding to a water/cement ratio of about 0.4. However, the strength of cement pastes increases as porosity decreases below this level, that is as the cement content of a paste increases even though some of the cement remains unhydrated, so that the gel space ratio concept is of limited applicability.

Fagerlund (1974) pointed out that Powers' Equation (8.7) is a special case of Bal'shins' equation:

$$\sigma = \sigma_0 (1 - \epsilon)^A \quad (8.8)$$

in which σ is the measured strength of the porous paste, σ_0 the strength of the solid phase, ϵ the porosity and A is a constant with a value close to 3.

Fagerlund found that different degrees of maturity required slightly different values for the constants in Equation (8.8) and that for pastes made with the same cement and the same water/cement ratio, strength up to 28 days was almost a linear function of porosity. He also observed that at high porosities data could be better fitted to the equation:

$$\sigma = \sigma_0 \left[1 - \left(\frac{\epsilon}{\epsilon_{cr}} \right)^{1/B} \right] \tag{8.9}$$

where B is an empirical constant and ϵ_{cr} a critical porosity, at which no strength is developed, having a value dependent on the volume of entrained air.

Other authors have fitted their results to an exponential equation. Roy and Gouda (1973) observed a good linear correlation between σ and log ϵ for very high strength pastes in which porosities from 20% down to about 2% were produced by hot pressing. These equations (which have also been used for ceramic materials), are empirical, that is derived from curve fitting and not deduced from a physical model. Powers considered the filling of space by gel to be the principal requirement in generating strength and Fagerlund pointed out that the way in which the hydration products interact (bonding and interlocking) must be of equal importance. The interplay of these factors can be expected to explain the subtle differences between cements with different compositions and between different methods of curing.

Plots of log σ against ϵ for samples of paste produced by different curing methods (at room temperature; in an autoclave; with and without added pulverised fuel ash) were found to be linear by Feldman and Beaudoin (1976). However, results could be grouped around three different straight lines as shown in Fig. 8.8, the large number of points they plotted being omitted for clarity. Results for pastes cured at room temperature with porosities ranging from 1.4 to 41.5% lay close to the line AB, while autoclaved preparations, excluding those made with an ash addition, followed the line OD with one result for a hot pressed sample with 6% porosity, near C. Those autoclaved pastes to which ash had been added gave results following the line EF.

Feldman and Beaudoin concluded from the different positions and slopes of the lines in Fig. 8.8 that strength depends not only on porosity but also on the proportions of coarsely crystalline and poorly crystalline material in a hardened paste. They suggested that the latter forms solid–solid bonds more readily and that this property is dominant above about 30% porosity, when room temperature pastes are stronger than the autoclaved pastes. Taylor (1977) provided further support for these ideas (Section 8.4).

It might also be considered that the most significant differences between the samples giving the results following the lines OD and OB in Fig. 8.8 reside in their pore size distributions. For the same total porosity,

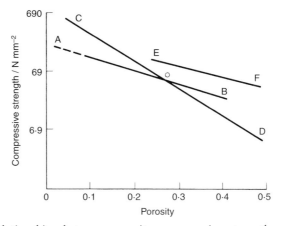

Fig. 8.8.   Relationships between porosity, compressive strength and curing of cement pastes (simplified diagram after Feldman and Beaudoin, 1976): AB cured at room temperature, CD autoclaved, EF autoclaved with added pulverised fuel ash

an autoclaved paste will contain far fewer small pores and more coarse pores than a normally cured paste, since thermal treatment involves an agglomeration of fine pores with a consequent decrease in specific surface area (Gregg and Sing, 1982). The coarsening of the pore system produced by autoclaving, which parallels the coarsening of the crystals of the hydration products, has been demonstrated (at least qualitatively) by mercury porosimetry and, for a given total porosity, compressive strength has been shown to increase with decreasing average pore radius (Jambor, 1973). Only the incorporation of ash, to react with the calcium hydroxide formed in hydration and increase the proportion of calcium silicate hydrates (Section 9.4.2), raised the strength of the products formed in the autoclave above those of the room temperature pastes at higher porosities (line EF in Fig. 8.8).

Strength is also dependent on the relative humidity with which the paste is in equilibrium when it is measured. An increase in flexural strength of about 50% was observed on first drying samples of hcp by Sereda *et al.* (1966) and a similar decrease on re-exposure to water vapour at 10% rh. A further increase in the rh to 98% had little effect. Pastes made at water/cement ratios of 0.3–0.6 responded in this way, but at 0.7 changes in strength with rh were slight. Wittmann (1973) reported similar trends in compressive strength and noted that the effect of rh was more pronounced on mortar and especially on concrete. It is not known whether the fall in strength when rh is raised is due to the effect of water vapour on the interparticle cohesive bonds in the gel or whether it enhances crack propagation by adsorption on newly formed surfaces.

8.3.3 Elastic and inelastic properties

Published plots of uniaxial compressive stress against strain for hcp are non-linear. The results in Fig. 8.9 illustrate the dependence of the relationship between stress and strain in hcp on the water/cement ratio and the maturity of the paste. The strain rate also influences the result obtained to a small extent. A dense aggregate gives a stress–strain plot which is much more nearly linear than that for a sample of hcp, and a concrete containing the aggregate gives a curve intermediate between the two. The determination of a uniaxial strain in hcp (or concrete) that is free from errors introduced by local distortions arising from platen-sample contact in the compression machine is difficult.

The curvature of the plots in Fig. 8.9 makes the characterisation of samples by a static elastic modulus somewhat arbitrary. The slopes of specific chords, or of tangents to the curve, may be quoted. Spooner et al. (1976), for example, gave an elastic modulus derived from the tangent to the initial stress–strain curve at a stress of $5 \, \text{N/mm}^2$. In general, the static elastic modulus for hcp increases with increasing compressive strength.

Dynamic elastic moduli have been determined from the resonant vibration frequencies of small discs and beams of hcp. Neville (1995) gives an introduction to the principles of the methods used and their application to the non-destructive testing of concrete. The dynamic modulus of a sample is higher than that obtained by so-called static or slow loading methods and, since the determination of resonant frequency can involve the application of very low stresses, values are believed to approach the initial tangent modulus.

Values of elastic moduli (E) for hcp obtained by static and dynamic methods have been shown to be dependent on the porosity of the paste. Helmuth and Turk (1966) fitted their results for both Portland cement and C_3S pastes to an empirical equation given by Powers (1961):

$$E = E_0(1 - \epsilon)^3 \tag{8.10}$$

where E_0 is a constant equal to the elastic modulus, either of the non-porous solid if total porosity (ϵ_T) is plotted, or of the gel if capillary porosity (ϵ_c) is employed. Plots of Young's moduli and shear moduli were found to be linear over a wide range of porosities (Fig. 8.10). Feldman and Sereda (1968), on the other hand, obtained a good fit of their static moduli to an exponential equation (Fig. 8.16).

Fagerlund (1974) pointed out that deviations from equations such as (8.10) must be expected in poorly cured pastes because E is strongly influenced by the shapes and interconnection of pores. He reviewed critically a number of relationships between porosity and elastic modulus and noted that E_0 values of about $8 \times 10^4 \, \text{N/mm}^2$ for cement hydrates, obtained by extrapolating plots of E against $(1 - \epsilon_T)^3$ to zero porosity, are

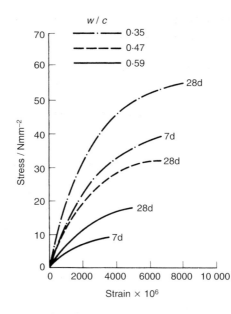

Fig. 8.9.   Stress–strain plots for cement pastes: effects of water/cement ratio (w/c) and curing (days) (Spooner, 1972), strain rate 300×10^{-6}/h

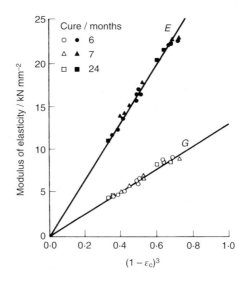

Fig. 8.10.   Relationship between dynamic moduli of elasticity and capillary porosity (ϵ_c) for hardened cement pastes (Helmuth and Turk, 1966); E Young's modulus, G Shear modulus

reasonable and similar to that of silica sand ($7 \times 10^4 \, \text{N/mm}^2$). E_0 for the solid phase in hcp is at least double that for the gel with its normal porosity.

By loading and unloading samples of hcp cyclically, Spooner *et al.* observed two types of hysteresis, that is energy loss. The first, found only in the first cycle, they ascribed to structural damage while the second, found in the first and also in subsequent cycles (in which the original maximum stress was not exceeded), indicated damping, that is an energy-dissipating process. They suggested that the latter might involve a sliding friction within the solid fraction of the paste. Damage occurred in the first cycle even at the lowest strains examined.

The elastic modulus of a gel can be regarded as determined by two components; the elasticity of the particles themselves and that of the three-dimensional network they form, the influence of the latter predominating at high porosities. Since the pore system is continuous, irreversible damage can be expected from the breaking of solid–solid bonds with a change in the particle packing in the gel. The above observations suggest that this can occur at low stresses in hcp. Reversible damping might be linked with more limited movement within the gel at the points of solid–solid contact.

Resonant frequency methods yield a value very close to the real part of the complex dynamic modulus E^*:

$$E^* = E_1 + iE_2 \tag{8.11}$$

A measure of the damping capacity of a material is given by the ratio of E_2 to E_1 and:

$$E_2/E_1 = \tan \delta \tag{8.12}$$

where δ is the phase difference (loss angle) between an applied sinusoidal stress and the resulting sinusoidal strain.

A plot of loss angle against the frequency of an applied sinusoidal stress exhibits a peak when the frequency coincides with the relaxation time of a process responsible for damping. Sellevold (1976) found a loss peak at about 0.1 Hz in a water-saturated hardened (autoclaved) cement paste. Sellevold and Richards had previously observed an early (< 10 s) transition in the time-dependent reduction in strain (*creep recovery*) when a similar sample of paste was unloaded. Sellevold showed that the loss peak at 0.1 Hz could be predicted from the creep-recovery response, on the assumption that hcp is a linear viscoelastic solid. Since the creep-recovery transition had been shown to have a temperature dependence similar to that of the viscous flow of water, this damping effect was attributed to the movement of water in the capillary pore system of the paste.

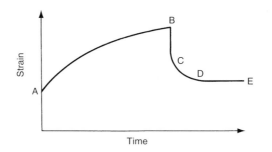

Fig. 8.11.   Recoverable and irrecoverable creep in hcp in hygral equilibrium: A — initial elastic strain, B — load removed, BC — elastic recovery, CD — creep recovery, E — residual deformation

All time-dependent, inelastic deformations accompanying sustained loading of a material are generally described as *basic creep* when the hygral state of the specimen remains unchanged. As shown schematically in Fig. 8.11, the total time-dependent strain is only partially recovered when the load is removed; of the strain which disappears, part does so instantaneously (BC — elastic recovery) and part slowly (CD — damped elastic or creep recovery). The recoverable part of creep (CD) reaches a maximum after a period of loading but the irrecoverable part continues to increase. An alternative manifestation of the processes involved in the creep is the time-dependent relaxation of stress in a sample of hcp held at constant strain. Such stress relaxation can be of value in reducing the cracking associated with drying shrinkage (Section 8.3.4).

If a sample of hcp dries while under compressive stress, the total of deformation is greater than that anticipated from the sum of the basic creep and drying shrinkage of an unstressed sample; the additional deformation is referred to as *drying creep*. The total (basic + drying) creep in concrete is an important practical parameter in structural engineering, although the magnitude of the effect is greatly reduced by the presence of a dense aggregate.

The rate of creep deformation is dependent on temperature and humidity. An idea of the rate of creep in hcp and the effect of relative humidity on it is given in Fig. 8.12 in which measurements have been fitted to a commonly used empirical power law:

$$C = at^n \tag{8.13}$$

where C is the creep deformation, t the time under constant stress and a and n are constants with values depending on the constitution of the paste and the conditions of test.

The mechanisms suggested as involved in creep are reviewed by Xi and Jennings (1992). A short-time reversible creep has been linked by

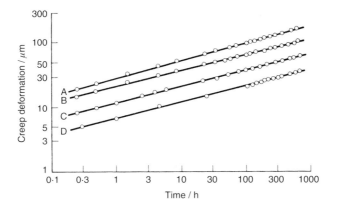

Fig. 8.12.   Creep deformation of hardened cement pastes as a function of time under load and equilibrium rh/%: A — 100, B — 98, C — 53, D — 11 (Wittmann, 1973)

Sellevold and Richards (1976) with the movement of water under stress gradients. At low stresses and for short times saturated hcp can be treated as a linear viscoelastic solid. It is likely that irrecoverable creep is linked with a rearrangement of the particle packing within the gel, damage of its structure, and perhaps also with the slow relative displacement of the layers within the C–S–H particles themselves (Feldman and Sereda, 1972). At least some of these inelastic processes are accelerated at high relative humidities and Wittmann (1970) suggested that this results from a weakening of the solid–solid bonds in the 'xerogel' (gel) by the swelling pressure introduced by a film of adsorbed water.

8.3.4   Drying shrinkage

The magnitude of the drying shrinkage of hcp makes this one of its most characteristic properties and one which must be considered in all engineering applications. Verbeck and Helmuth found a total linear drying shrinkage of 1.2% with a sample of hcp (water/cement ratio 0.5), just over half of which occurred below about 40% rh, although only about a quarter of the total amount of water removed by drying was lost at this stage. This suggests a change in shrinkage mechanism. For mature pastes, drying shrinkage increases with increasing water/cement ratio. In structural concrete, shrinkage is restrained by a high content of non-porous aggregate. For concrete with a 70% volume content of aggregate and unhydrated cement, shrinkage is only about one-fifth of that of cement paste.

Helmuth and Turk (1967) found that hcp exhibited a large first drying shrinkage, not fully recovered on resaturation, and a reversible subsequent drying shrinkage. It is seen in Fig. 8.13 that only the irrecoverable

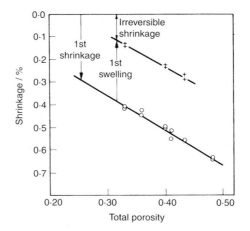

Fig. 8.13. Relationship between drying shrinkage and total porosity for hardened cement pastes (Helmuth and Turk, 1967)

component is dependent on the total porosity of the paste. They also observed that the first drying shrinkage increased with time when the paste was held on 47% rh before completion of the drying.

The total shrinkage of a paste increases with increasing degree of hydration, indicating a link with the content of hydration products. Although there are some differences between results reported for various forms of drying in the literature, the general picture is that structural changes which are dependent on time, temperature and humidity occur on first drying, so that the gel structure becomes more stable, with smaller dimensional changes in subsequent wetting–drying cycles.

Drying shrinkage has been linked with water adsorption as a function of relative humidity by Feldman and Sereda (1968), using the model for cement paste shown in Fig. 8.14. In this model C–S–H particles are considered to possess a distorted layer structure (B) with water molecules in interlayer sites (X). The number of layers per particle is small and variable in accord with the fact that no long-range order is revealed by X-ray diffraction. Individual C–S–H particles are linked in a three-dimensional structure by solid–solid bonding (A).

Relationships between the adsorption isotherm and changes in length are indicated schematically in Fig. 8.6. In the region OA physical adsorption and interlayer rehydration of C–S–H particles are both considered to occur. In this region the Gibbs adsorption isotherm (Equation (8.14)) can be used but only for thermodynamically reversible adsorption. This equation links the amount adsorbed (n mol/g solid) with water vapour pressure (p_w) and with the change in the surface tension ($\Delta\gamma$) of the solid produced by the adsorption:

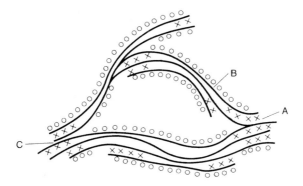

Fig. 8.14. *Feldman–Sereda model of hardened cement paste — a network of the poorly ordered particles forms a porous three-dimensional array: A — solid–solid interparticle bond, B — C–S–H layers within a particle, C — narrow entrance to a gel pore, X — interlayer water, O — physically adsorbed water*

$$\pi = \Delta\gamma = RT/S \int_{o}^{p_{\mathrm{w}}} \frac{n}{p_{\mathrm{w}}} \, dp_{\mathrm{w}} \tag{8.14}$$

where S is the specific surface area, R and T have the usual meaning and π is the surface (spreading) pressure of the adsorbed film, which is equal to its partial molar free energy $(\partial G/\partial A)_{T,p,n}$.

The surface atoms or ions of a solid have a coordination number which is only a fraction of that for atoms in the bulk of the solid. This results in the equivalent of a hydrostatic pressure (P) on a particle which is related to its radius r, assuming it is spherical, by the Laplace equation:

$$P = 2\gamma/r \tag{8.15}$$

For a particle of C–S–H with a radius of 10 nm, P is of the order of 2500 atmospheres (Wittmann, 1968). Adsorption on the surface when the solid is exposed to a gas reduces the imbalance of forces on the surface atoms and the particle swells.

The Bangham equation (8.16) can be used to relate fractional length change $(\Delta L/L)$ to the changes in surface tension and it is applicable up to about $0.45P_{o}$. For small values of ΔL the volumetric swelling is $3\Delta L$:

$$\Delta L/L = k\Delta\gamma \tag{8.16}$$

Feldman (1968) found that for the adsorption of water on hcp, a plot of $\Delta L/L$ against $\Delta\gamma$ (determined from the measured adsorption by applying the Gibbs adsorption isotherm) yielded a straight line through the origin only when the thermodynamically reversible physical adsorption was used to calculate $\Delta\gamma$. He derived the reversible adsorption from a series of adsorption–desorption cycles which made it possible to determine the irreversible adsorption occurring at the interlayer sites and deduct it from the total adsorption after each incremental increase in relative pressure. It must be noted, however, that other workers have not used such a procedure and general agreement on the role of interlayer rehydration in the swelling of hcp has not been reached.

Capillary condensation occurs at higher relative pressures (region AB in Fig. 8.6) and in the region CD dimensional changes in a porous solid are caused by changes in the tension in the capillary condensed water accompanying changes in the radius of its meniscus. As the vapour pressure is reduced, the radius of the meniscus of the water in a cylindrical pore decreases from infinity at saturation (completely full) to a minimum radius equal to the pore radius. If there is any further reduction in vapour pressure the pore empties and the quantities involved (pore radius and relative pressure) are linked by the Kelvin equation (Appendix). The tensile stress in the water is balanced by compressive stresses in the solid so that as the former increases, with decreasing relative pressure along CD, the solid shrinks. In the region DA, pore emptying accompanied by swelling is predominant and when, as is frequently the case, this occurs at $0.4–0.45p_o$, it may be due to the fact that the tension in the capillary water then exceeds its tensile strength ($\sim 1.2\,kN/mm^2$) or to the presence of ink-bottle pores, that is wide pores with narrow entrances (Appendix).

Another cause of volume changes at relative humidities greater than about 0.45, developed by Powers (1969), is considered to involve the free energy of adsorption of water as its driving force and the fact that there are areas of 'hindered adsorption' in the paste. Thus, the thickening of the adsorbed film of water as relative humidity increases is possible at X in Fig. 8.15 if the gel particles are forced apart so that the paste swells. The term 'disjoining pressure' (Derjaguin, 1935) is often used when this swelling mechanism is being described but it is not always made clear that this is a property of a liquid film on a surface and not another name for the swelling process.

The concept of hindered adsorption as a cause of swelling is not universally accepted but the process of *intercalation*, which involves the interlayer penetration of an adsorbent by an adsorbed film, is known to occur, usually at high relative pressures, with a number of adsorbents (Everett, 1967). In a particulate solid there will be competition between interparticle adhesive forces, as well as intraparticle cohesive forces, and the spreading pressure of an adsorbed film, which is related to the relative pressure of the adsorbate by the Gibbs equation (8.14). It is possible for the separation of two solid surfaces in contact to occur if the decrease in free energy resulting from increased adsorption (equal to $\pi\,dA$) is greater than the increase brought about by destroying the adhesive or cohesive bonds over the same area (Fig. 8.15).

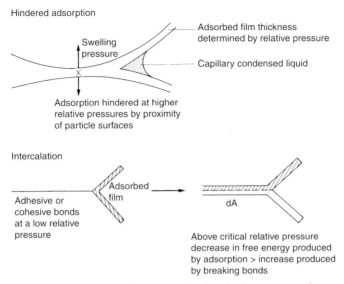

Fig. 8.15. Schematic representations of hindered adsorption and intercalation

The interaction of water with swelling clays is a special case because of the significant contribution to the energetics of the process made by interlayer cation hydration. Interlayer bonding in a clay crystal is uniform, so that water penetration occurs when a threshold spreading pressure is reached, giving a sudden increase in adsorption, and Van Olphen (1965) found a stepped isotherm for water adsorption on sodium montmorillonite. No similar distinctive step has been found in an isotherm for hcp but if there is an analogous process in C–S–H, then because the solid–solid adhesion and the interlayer cohesion must be very variable in such a material, the existence of a critical relative humidity would not be anticipated. The steep part of the isotherm at low pressures (Fig. 8.6) results from the relatively strong interaction between water molecules and the outgassed surface to form a monolayer and this would mask intercalation at this stage. The fact that hcp retains its integrity at the saturated vapour pressure of water indicates that at least some of the solid–solid contacts in the gel undergo only limited, if any, penetration. If this were not so, hcp would be dispersible!

Usually failure to remove all the adsorbate when the relative pressure is reduced to zero has been taken as evidence for intercalation. However, Brunauer and co-workers found that by considerably extending equilibration times in determining the desorption branch of the water isotherm for hcp, a closed hysteresis loop was obtained (dashed line in Fig. 8.6). Evidence for intercalation in C–S–H has come from changes in the rate of penetration of helium into hcp as small increments of water are adsorbed and desorbed. The results of such studies (Feldman, 1972) indicate that interlayer water in C–S–H is removed by drying below 11% rh and that re-entry into interlayer sites occurs at low rh, accompanying the adsorption of a monolayer on the free surface (sites o in Fig. 8.14).

The Feldman-Sereda model of hcp is only one of a number of models published which seek to relate the microstructure of a paste with its physical properties and the effects of rh and stress on them. Xi and Jennings (1992), in a critical review of relationships between the microstructure, creep and drying shrinkage of hcp, discussed the models which have been proposed. They concluded that, while attention has mainly been focused on the effects of rh on C–S–H, a model yielding quantitative relationships widely applicable to hcp must include the restraining effect of the crystalline components, such as CH and residual clinker. The presence of these phases, which are relatively unaffected by changes in rh, means that, like concrete, hcp must be modelled as a composite material.

8.4   Nature of the solid–solid bond in hardened cement paste

The ease with which Portland cement functions over a wider range of water/cement ratios than is possible for systems in which the products of hydration are well developed crystals, suggests that a colloidal product such as C–S–H has a special function related to the solid–solid bonds in hcp. This idea is supported by the work of Feldman and Sereda (1968) which showed (Fig. 8.16) that elastic moduli similar to those of hcp could be obtained by compacting hydration products prepared in powder form by using a water/cement ratio of about 10, a process referred to as *bottle hydration*. In contrast, gypsum crystals obtained by hydrating plaster at a high water/solid ratio did not yield compacts with elastic properties similar to cast plaster.

Taylor (1977), following Feldman and Beaudoin (Section 8.3.2), related the strength developed by cement to porosity and the proportion of coarse crystalline material formed in hydration, indicating trends in strength by contours (Fig. 8.17). The lines AB, CD and EF correspond to those similarly labelled in Fig. 8.8. It is seen that to obtain high strength with a high content of coarse, crystalline hydration products (produced in the autoclave, for example) a very low porosity is necessary. With a high proportion of colloidal hydrates, the paste is weaker at low porosities but more tolerant of higher porosities. The broken line XC in Fig. 8.17 joins points of maximum strength (optimum composition) at a given porosity.

The actual nature of the solid–solid bond in hcp is still subject to speculation. Secondary valency bonds, such as Van der Waals' forces and hydrogen bonding involving water, have often been suggested. At the

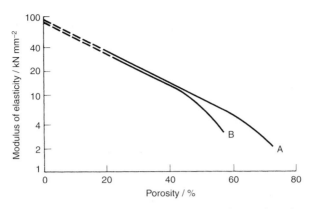

Fig. 8.16.   Comparison of elastic modulus–porosity relationships for pastes (A) and for samples (B) obtained by compaction of hydration products obtained in powder form by bottle hydration (Feldman and Sereda, 1968)

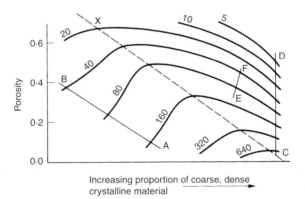

Increasing proportion of coarse, dense crystalline material

Fig. 8.17.   Relationships between porosity, crystallinity and strength of pastes (Taylor, 1977). Contours connect points of equal strength (N/mm^2); letters have the same significance as in Fig. 8.8

high pH existing in cement hydration and in the presence of the dissolved calcium ions, both silica- and alumina-rich surface sites will be negatively charged with calcium counterions. The latter might be expected to behave on drying as they do in a swelling clay, that is to provide ionic bonding to link surfaces. However, the effectiveness of such bonding may be very limited, since it depends on the density of the surface charge and the closeness of approach of the solid surfaces on drying. Clay minerals are a special case since flat surfaces with a significant charge density, defined by isomorphous solid solution, are involved.

Primary, chemical bonding involving the development of Si–O–Si links between particles in mature pastes has also been suggested. If such primary bonding developed to a significant extent then hcp would be expected to have a higher tensile strength, but this must depend as much on the total area of solid contact as on the bond type. In this respect, the deformability of colloidal, disordered particles such as those of C–S–H gives them an advantage, suggested by the results in Figs 8.8 and 8.16, in establishing cementitious contacts, at least at higher paste porosities.

Although Mössbauer spectroscopy does not indicate the nature of the interparticle bonds in a gel, it has provided a technique for the estimation of their strength. The method depends on the fact that the high energy associated with the emission of a γ photon from the ^{57}Fe nucleus produces a recoil which is damped to an extent depending on the rigidity and mass of the solid in which the nucleus is bound. For particles of colloidal dimensions that is no Mössbauer resonance but if they are linked in a gel then broad peaks are obtained and their area can be used to estimate the strength of the interparticle bonding. Ubelhack (1976) applied the method to a paste of calcium aluminate cement (Section 10.3), chosen for its suitable iron content, and related interparticle bonding in the 'xerogel' to its elastic modulus.

8.5 Modelling microstructure–physical property development in hydrating hcp

The increasing use of a computer to model changes in the microstructure of a cement paste, as hydration proceeds according to an accepted rate equation, has been an interesting recent development. By incorporating a link with physical properties such as compressive strength, the desired outcome is the substitution of repetitive testing by computed physical properties as a function of curing. The modelling process is one of space filling, starting with an array of clinker particles with the appropriate particle size distribution. These are then 'hydrated' using an algorithm, allowing for dissolution and diffusion of ions, based on rules selected as a result of fundamental studies of the hydration of the constituent phases of cement. Curing time is represented by iterations of the algorithm so that space in the 'map' occupied initially by water is progressively replaced by hydration products. A useful introduction to computer modelling has been given by Garboczi and Bentz (1991).

Jennings and Tennis (1994) used computer modelling to predict the development in a maturing paste of the volumes of C–S–H, AFm, CH, unreacted clinker and pores. As pointed out in Section 8.2.1 their model included two forms of C–S–H. They used constants for the Avrami-type equations deduced by Taylor (1984) for each of the clinker constituents to introduce rates of hydration. Their model predicted a surface area of $392 \, \text{m}^2/\text{g}$ for D-dried C–S–H, comparable with the 350–$450 \, \text{m}^2/\text{g}$ obtained by small angle X-ray scattering and adsorption isotherms.

Bentz (1997) described the development of a three-dimensional model, pointing out that while phase volumes could be statistically represented in two dimensions, pore connectivity which determines permeability can not. He found the Knudsen parabolic law (Section 7.5) held for the hydration of his cement samples and used the Powers-Brownyard gel space ratio concept in calculating strength development. Measured mortar strengths correlated well with predicted strength–time plots. The paper contains a detailed account of methodology and a review of the earlier literature of modelling.

8.6   Concluding remarks

Hardened Portland cement paste is a gel with a high solids content, some of it crystalline, and a considerable insight into its microstructure has been gained from the application of electron microscopy. A number of models have been proposed for hcp which attempt to correlate its (macro) physical properties with the characteristics of the gel which are: a high specific surface, located in a complex pore system ranging from micropores up to capillary pores; a range of sites for water vapour adsorption and significant volume changes (reversible and irreversible) linked to changes in relative humidity; solid–solid bonding which limits swelling at high relative humidity; susceptibility to undergo significant creep deformation when subjected to sustained stress. Models proposed have generally focused on the structure of C–S–H and the need to develop a model of hcp microstructure as a composite material, containing crystalline components relatively unaffected by changes in rh, was emphasised by Xi and Jennings.

Like other polycrystalline ceramic materials, the strength of hcp can be quantitatively (although empirically) linked to its total porosity but there is strong evidence that the poorly crystalline components in it, because of their adaptability in filling space, have a special role in solid–solid bonding and greatly enhance strength at higher porosities.

Although the flexural strength of hcp is normally of the order of only $10\,\text{N/mm}^2$, Birchall et al. (1983), recognising the importance of pore size distribution, prepared pastes of both Portland and calcium aluminate cements free of large pores and with flexural strengths in excess of $100\,\text{N/mm}^2$. They employed selected particle size fractions and water/ cement ratios less than 0.2, using high molecular weight polymer additives to make the mix cohesive. The flexural strengths reached are comparable to those of dense, sintered ceramics and like these the hardened paste is brittle. Optimum properties were achieved with calcium aluminate cements (Section 10.3). The term 'macro defect free cement' was introduced but this is a description of the hardened paste and not of the cement itself. They are now regarded as more accurately described as cement-polymer composites with ions from the cement providing possible cross-linking. Unfortunately these products, in spite of their very low permeability, proved to be dimensionally unstable in wet conditions, but interest continues and other such composites based on calcium sulfoaluminate cements (Section 10.5) remain under investigation. A summary of the properties and potential of these materials was given by Young (1993).

In the past 15 years, increased awareness of the influence of portlandite on the pore size distribution of concrete has provided the background to the development of high strength, low permeability concretes employing pozzolanas, metallurgical slags and silica fume. These materials, which can significantly reduce the portlandite content of a hardened paste, are considered in Chapter 9.

9. Portland cements and related blended cements

9.1 Introduction

The European (voluntary) prestandard specification, ENV 197-1: 1992, groups cements commonly available in Europe into five types. So far this book has dealt primarily with Type I defined, in the terminology adopted by the standard, as made up of a *nucleus* consisting of 95–100% of Portland clinker plus gypsum (usually about 5–6% depending on purity) to the required SO_3 content and perhaps 0.1% of a grinding aid. Up to 5% of a minor additional constituent is permitted as part of the nucleus (Section 5.1.2). In the past, BS 12 has specified 100% clinker in Portland cement for it to comply with the standard but the revision in 1991 paralleled the then existing draft of ENV 197 in allowing a minor additional constituent, such as limestone or one of the materials considered in Section 9.3. An abbreviated summary of the cement types designated in ENV 197-1 and existing relevant British Standards is given in Table 9.1. The ranges of cement compositions defined in the European standard are required to cover the wide range of products commercially available in the Member States.

9.2 Type I and related Portland cements

Table 9.2 lists the available Type I cements using the names traditionally employed in the UK. The designations of the nearest ASTM equivalents are included as they are frequently encountered in the technical literature. The table also lists closely related Portland cements (5–7) not included in ENV 197-1. They will be covered by parts of the standard still under development. Number 8 is a Portland cement composition available in the USA which is included in the ASTM classification. It is 'modified' in having a moderate heat of hydration and moderate sulfate resistance as a result of a limited C_3A content (Section 9.2.2).

Table 9.1.   Classification of European cements (ENV 197–1: 1992)

Cement type	Designation	Notation	Constituents		Corresponding BS
			Clinker: %	Other main: %	
I	Portland cement	I	95–100	—	BS 12: 1996
II	Portland:				
	silica fume	II/A–D	90–94	6–10	—
	slag	II/A–S	80–94	6–20	BS 146: 1996
		II/B–S	65–79	21–35	
	pozzolana	II/A(B)–P	Composition ranges A and B		—
		II/A(B)–Q	as for Portland slag cement		—
	fly ash	II/A(B)–V			BS 6588: 1996
		II/A(B)–W			—
	burnt shale	II/A(B)–T			—
	limestone	II/A–L			BS 7583: 1996
		II/B–L			—
	composite	II/A(B)–M			—
III	Blastfurnace cement	III/A	35–64	36–65	BS 4246: 1996*
		III/B	20–34	66–80	—
		III/C	5–19	81–95	—
IV	Pozzolanic cement[†]	IV/A	65–89	11–35	—
		IV/B	45–64	36–55	BS 6610: 1996
V	Composite cement[‡]	V/A	40–64	36–60	—
		V/B	20–39	61–80	—

* Covers blastfurnace slag content of 50–85%.
[†] One or more of: silica fume, natural/industrial pozzolana, siliceous fly ash.
[‡] Blastfurnace slag + one or more of: natural/industrial pozzolana, siliceous fly ash.

The pre-standard defines a cement nucleus as: clinker + other main constituent(s) (if any) + minor additional constituent (0–5% permitted, Section 5.1.2) expressed as mass percent to total 100.

The cement itself is defined as a nucleus + gypsum (max. determined by SO_3 limit) + additive (0–1%) — usually a grinding aid at ca. 0.1%.

Designation of other main constituents: D silica fume; S granulated blastfurnace slag; P natural pozzolana; Q industrial pozzolana; V siliceous fly ash; W calcareous fly ash; T burnt shale; L limestone; M mixture of permitted main constituents.

9.2.1   Cements covered by BS 12: 1996

Rapid hardening cement is similar to the ordinary variety, which usually has an adequate C_3S content, except that it is more finely ground. Unlike ENV 197, BS 12: 1996 does not allow more than 3.5% SO_3 in higher strength classes. The increase in early reactivity which results from the higher specific surface increases the heat of hydration at early ages making this cement unsuitable for large structures where heat cannot be readily dissipated. It is, however, of value in cold weather concreting.

White cement is manufactured from the purest raw materials naturally available. In addition to an iron oxide content of less than about 0.3%, the

Table 9.2.   Type I   l Portland cements

Portland cement	ristics	British Standard	ASTM* Type
1. Ordinary[†]	sp. surf. 330–380 m²/kg strength class 42.5 N $SO_3 \not> 3.5\%$[‡]	BS 12: 1996	I
2. Rapid hardening	sp. surf. 400–450 m²/kg strength class 42.5R or 52.5 N	BS 12	III
3. White	$Fe_2O_3 < 0.3\%$	BS 12	—
4. Controlled fineness	sp. surf. agreed with customer	BS 12	—
5. Low heat	heat hyd. < 250 j/g at 7 d < 290 j/g at 28 d[‡] sp. surf. > 275 m²/kg[‡]	BS 1370: 1979	IV
6. Low alkali	$Na_2O\% + 62/94K_2O\% < 0.6\%$	(BS 4027: 1996)	—
7. Sulfate-resisting	$C_3A \not> 3.5\%$[‡] $SO_3 \not> 2.5\%$[‡]	BS 4027: 1996	V
8. Modified	$C_3A \not> 8\%$[‡]	—	II

* Nearest equivalent in the ASTM designation.
† The adjective 'ordinary' no longer has the status of a British Standard classification.
‡ Requirement of a standard.

content of manganese oxide must be one order and of chromium oxide at least two orders of magnitude lower than this. The usual raw materials are a china clay (kaolinite) and a selected white chalk or limestone. If these do not contain the additional free silica needed to produce the required level of silicates, then ground white sand is added to the raw materials mix. The particle size range of the silica and the content and aluminous nature of the flux determine the required burning temperature, which may be in excess of 1600°. For optimum whiteness it is also usually necessary to burn under slightly reducing conditions and cool the clinker by rapid quenching to counter the effects of manganese and chromium, since even very low levels impart a bluish-green tinge. Together, these factors ensure that white cement is significantly more expensive than ordinary Portland cement and it is used mainly for special architectural purposes such as rendering.

Controlled fineness cement is produced in Britain on a small scale by grinding to a fineness meeting the needs of a particular industry. For example, in the manufacture of cement fibreboard it is necessary to dewater a slurry of the cement and the fibre. To facilitate this, a coarsely ground cement is employed.

Low heat cement was considered in Section 6.7. Although it is not now manufactured in Britain a standard, BS 1370: 1979, exists.

Low alkali cement. In the USA since the 1940s, and Denmark since the 1950s, it has been known that a significant proportion of the siliceous aggregate available is susceptible to attack by the alkaline medium produced by cement, where concrete is exposed to wet conditions. However, it was not until 1976 that it was established that cracking due to alkali-silica reaction (ASR) had occurred in UK maintained structures. The number of cases found in Britain increased significantly during the 1980s.

Reaction of certain forms of silica in an aggregate, opaline silica or a volcanic glass for example, leads to the formation at its interface with the cement paste of an alkali-silicate gel which swells by absorption of water (pore solution). This is a process of osmosis involving the hydration of cations fixed within the gel into which water diffuses. It is analogous to the process of swelling of smectite clays described by van Olphen (1977). Swelling can result in expansion, cracking and deterioration of the concrete in a manner dependent on its mass, restraint such as reinforcement and additional stresses such as those induced by wetting and drying cycles. Deterioration due to ASR may take some time to appear and a silicate gel may eventually diffuse through the cracks to the concrete surface and, as a result of carbonation, produce a white deposit of silica gel and calcium and alkali carbonates. The role of the alkali metal ions is primarily one of enhancing the concentration of OH^- ions in the pore water of the wet concrete (Section 7.3), to which is added the concentrating effect of increased binding of water as hydration progresses. However, it should be noted that the pH of the pore solution produced in the hydration of Portland cement is also affected by the concentration of other anions present, such as sulfate, since they will also be involved in the cation-anion charge balance.

The alkali-silica reaction can be followed in laboratory specimens by means of expansion measurements on mortar, or preferably concrete specimens with proportions simulating field concrete. Of particular note is the fact that, with chosen mix compositions, the measured expansion may increase to a maximum with an increasing proportion of reactive silica in the aggregate and then fall again to zero with a further increase (Hobbs, 1981). This pessimum in reactive silica level appears to result from the fact that the key to this reaction is the concentration of alkali in the pore solution per unit mass of accessible reactive silica. In a well-mixed concrete the supply of Ca^{2+} at the reacting interface is maintained by the presence of calcium hydroxide formed by hydrating cement. Consequently, increasing the surface of reactive silica leads to an increase in the formation of non-swelling C–S–H with adsorbed alkali ions rather than a swelling alkali-rich silicate (Powers and Steinour, 1955). The reaction then is analogous to a pozzolanic one (Section 9.4). A decrease in expansion of a mortar also results if, at a given content, a reactive siliceous aggregate is finely ground.

NMR spectra, enhanced by using silica enriched in ^{29}Si, have elucidated some aspects of the alkali-silica reaction (Wieker, 1997). Examination of the course of chemical reactions of enriched silica with portlandite, alkali and C_3S in different combinations has indicated that at high alkali/lime ratios silica solubility is raised sufficiently to prevent the precipitation of C–S–H and a substance with a layer-silicate anion (revealed by a peak with a characteristic chemical shift in the NMR spectrum) is formed. This is considered to be analogous to the mineral kanemite $Na(HSi_2O_5).2H_2O$ which is a swelling silicate, a property retained after the incorporation of some calcium by ion exchange.

In field concrete, characteristic patterns of surface cracking and sometimes white, carbonated exuded gel may appear. Idorn *et al.* (1992) reviewed the phenomenon and test methods developed by ASTM for assessing the alkali sensitivity of aggregates. Mortar bar expansion methods are slow and accelerated tests have been proposed in which the reactivity of an aggregate with caustic soda solution is examined. However, while it is possible to identify a reactive aggregate in a laboratory test, there is no simple link between the result and the probability of excessive expansion when the aggregate is in contact with cement (Neville, 1995). Reactivity is linked with the completeness of the condensation of silanol groups in a mineral, ranging from fully condensed, unreactive crystalline quartz to disordered, porous minerals such as opal, which contain uncondensed hydroxyl groups. It is important to note that most as-dug aggregates contain a mixture of minerals, and quartz is potentially reactive when its crystals are strained. Its polymorph, cristobalite, is moderately reactive and was proposed as a replacement for opal as a reference aggregate in assessing the aggressiveness of cements (Lumley, 1989).

With a reactive aggregate, therefore, a low alkali cement is required in order that the pore water does not become too aggressive, and past experience had suggested that a total alkali content of less than 0.6% 'soda molar equivalent' (i.e. $Na_2O\% + 62/94\ K_2O\%$) was acceptable. However, the amount and concentration of alkali in a concrete will depend on the mix proportions and sources of alkali other than the cement and so a limit per cubic metre of concrete is preferred. Hobbs (1978) proposed a maximum of 3 kg Na_2O equivalent/m^3. Recent guidance for concrete production (Building Research Establishment Digest 330) recommends alkali levels from 5.0 down to 2.5 kg/m^3 depending on the reactivity of the aggregate.

The alkali content of a cement is determined by the alkali contents of the raw materials and fuel employed in its manufacture and the proportion retained in the clinker in the burning zone. It is sometimes possible to use a low alkali fly ash as an argillaceous component in the kiln feed. In the now predominant dry process a significant reduction in thermal efficiency would be involved if it were necessary to bleed off much of the exhaust gases leaving the kiln to prevent recycling of alkali-rich dust, which is trapped in the preheater and returned to the kiln. In Britain the reduced levels of alkali-containing clays or shales used in the manufacture of sulfate-resisting Portland cement make it suitable for use where a low alkali cement is required and this is allowed for in BS 4027: 1996.

9.2.2   Sulfate-resisting Portland cement (SRPC)

9.2.2.1   Sulfate attack.   If concrete is placed in soils containing sulfates or if it is in contact with seawater, then a sulfate-resisting cement must be used. The sulfates usually encountered naturally are those of sodium, potassium, calcium and magnesium. Such attack may also occur as a result of contact with effluents or contaminated soils. Sulfides of the heavy metals such as iron pyrites occur naturally in some clays and shales and oxidise when exposed to air to produce sulfates. The acid resulting from the accompanying hydrolysis of them adds to the aggressiveness of the environment. On the alkaline side of neutrality the aggressiveness of a sulfate solution falls as the pH increases up to 12.5, the value for saturated calcium hydroxide. In practice, of course, the pH of most natural groundwaters lies in the range 5.5–8.5 depending on rock/soil type. Penetration of sulfate ions into concrete made with ordinary Portland cement results in expansion with cracking and spalling, initially at edges and corners, and reduction in strength results. If the supply of sulfate ions in the groundwater is sufficient to continually replenish their concentration at the surface of the concrete, then ultimately complete softening and disintegration result.

In the laboratory, the effects of immersing concrete, mortar or paste samples in a sulfate solution can be followed by measurement of strength reduction or the usual accompanying expansion. The latter, being non-destructive, is favoured since far fewer samples are required, although reduction in compressive strength has been found to be the more sensitive indicator (Mehta, 1992). Lawrence (1992) noted the value of visual assessment of superficial damage to specimens as an early indication of performance. This method of assessment had been employed in extensive trials of cements by the Building Research Establishment in the UK and the Portland Cement Association in the USA.

The ASTM C1012 procedure involving measurement of the expansion of mortar bars in a solution of a sulfate (renewed at prescribed intervals) is frequently used. A 10% solution of sodium sulfate is often employed but testing with the more aggressive magnesium sulfate and a range of concentrations of solutions of each should be included. In this test specification an expansion of 0.5% is considered as a maximum acceptable limit. Assessment on a pass/fail basis is made difficult because of sensitivity of the result to aspects of the test procedure such as specimen size, compaction and cure before exposure (Frearson and Higgins, 1988). ASTM C1012 specifies cure of a mortar to a minimum strength of $20\,N/mm^2$ before immersion in the sulfate solution. Accelerated tests employing concentrated sulfate solutions are considered only to be acceptable when used for comparison of cementitious binders of the same type (Lawrence, 1992).

Since the nineteenth century it has been known that damage to concrete by sulfate solutions is accompanied by the formation of ettringite (Idorn *et al.*, 1992). As pointed out in Section 6.6, it is the delayed, expansive formation of solid phases such as sulfoaluminates in an already hardened structure with limited elasticity, which is linked to destructive effects. However, the *primary cause* of expansion and cracking remains the subject of debate. Some authors favour the crystal growth pressure theory which holds that expansive pressure develops at ettringite ($C_6A\bar{S}_3H_{32}$) crystallises from a supersaturated solution formed by the interaction of penetrating sulfate ions with either calcium aluminate hydrate (C_4AH_{13}), the monosulfate ($C_4A\bar{S}H_{12}$) or a solid solution of the two. The interconversion of these last two phases can occur by simple sulfate-hydroxyl ion exchange, that is topochemically, but the major structural reorganisation involved in their conversion to ettringite must involve a through-solution process. The reaction (Equation (9.1)) is accompanied by a significant increase in solids volume but, like the hydration of free lime (Section 6.6), a reduction in the total volume of solids and water.

$$C_3A.CaSO_4.12H_2O + 2Ca^{2+} + 2SO_4^{2-} + 20H_2O \longrightarrow$$
$$C_3A.3CaSO_4.32H_2O \tag{9.1}$$

Mehta (1973) suggested that swelling results from the imbibition of water by freshly formed colloidal ettringite. This would involve a combination of capillarity and osmotic effects. Odler and Gasser (1988) found that expansion could occur without water uptake by a specimen but that when it did occur, the expansion was significantly greater. It seems probable that the expansion is dependent on the pore size distribution in the specimen concrete or mortar since this will determine how far water already contained within the framework will sustain reaction (9.1) and to what extent, therefore, external water must be drawn in for it to proceed.

A number of other damaging processes may exacerbate sulfate attack on concrete (Neville, 1995). These include: salt penetration accompanying wetting and subsequent crystallisation on drying; freeze-thaw cycling; carbonation; and alkali-silica reaction. In seawater the tidal and splash zones are particularly susceptible to these processes and to them can be added abrasion by suspended sand. Mehta (1992) pointed out that much field experience suggests that a reduction of the alkalinity of hardened cement paste by magnesium hydroxide precipitation and carbonation by dissolved carbon dioxide are the primary causes of concrete deterioration in seawater rather than ettringite formation. At low temperatures, carbonation in the presence of limestone as a filler or aggregate can lead to the formation of thaumasite in sulfate attack on concrete (Crammond and Halliwell, 1997). Thaumasite ($Ca_3Si(OH)_6$ $12H_2O.SO_4.CO_3$) is unusual in containing octahedrally coordinated silicon without requiring high pressure to stabilise it.

Part of the problem of identifying the primary cause of damage associated with sulfate penetration is that chemical and physical changes progress slowly both with time and with distance from the surface at which water and ions enter the concrete or mortar. Consequently, finding a cluster of crystals in a sectioned specimen cannot be taken as indicating that their growth there is the primary cause of swelling and cracking, as they may have crystallised from a solution which had diffused into a preformed crack. Several authors have commented on the absence of proportionality between the degree of swelling observed with a laboratory mortar specimen and the amount of ettringite found in it. One factor limiting stress development resulting from reaction (9.1) is that continuing hydration will involve the dissolution of cement constituents which will reduce stress build-up. A second factor is that once initial swelling has induced cracking, additional pore space is created in which further ettringite crystals can grow without adding to localised expansive stress. The appearance of ettringite with two distinctly different crystal sizes (finely dispersed and more massive aggregates) is indicative of this possibility.

Gollop and Taylor (1992) examined the effect of 0.3M sodium and magnesium sulfate solutions on ordinary Portland cement paste (0.3 water/cement ratio, wet cured for 7 days). Polished sections taken at different depths of penetration and unaffected cores of the attacked specimens were examined by BSE imaging and microanalysis in the scanning electron microscope. In this way changes as a function of both distance and level of sulfate penetration could be observed. In a zone adjacent to the unaffected core of a sample the monosulfate had been replaced by ettringite, although the presence of both was only indirectly revealed by microanalysis as they were dispersed in the C–S–H gel at submicron level. In a highly cracked zone near the surface, gypsum was detected, both dispersed and as a massive form in veins (Fig. 9.1), which they considered to have resulted from diffusion and crystal growth in preformed cracks.

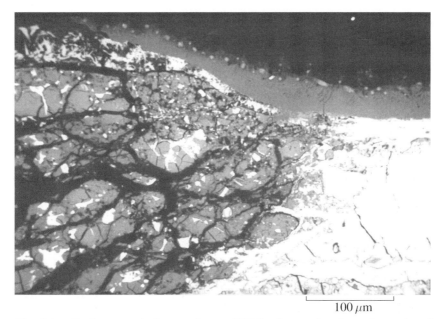

100 μm

Fig. 9.1.   Backscattered electron image (SEM) of a section through a cube of hardened cement paste which had been stored in magnesium sulfate solution. The mid-grey band at the face of the specimen is brucite with some aragonite at its outer surface. The corner of the specimen (top left) has largely been converted to a hydrated magnesium silicate with many cracks filled with resin (black). Below the surface on the right there are veins of gypsum (light grey) and unhydrated clinker (white) (courtesy of R.S. Gollop)

Gollop and Taylor emphasised the fact that the conversion of monosulfate to ettringite requires additional calcium ions (Equation (9.1)) and that they must be derived from the gel matrix. Microanalyses indicated that, initially, the calcium was derived from the calcium hydroxide component but subsequently from C–S–H. Such decalcification will contribute significantly to observed strength reduction. It should also be noted that for complete reaction 287 mg of water are required for every gram of monosulfate converted to ettringite. Sodium ions from the sulfate solution were not found to penetrate an initially undamaged paste and an increase in the pH of the solution indicated that ion exchange ($2OH^-$ for SO_4^{2-}) was involved. In the case of magnesium sulfate, precipitation of the relatively insoluble hydroxide also occurred at the surface of the paste and this buffered the aqueous solution at a pH of around 8–9. At this pH, C–S–H is unstable and extensive damage to the paste will occur. Lawrence (1992) pointed out that this pH range provides an environment closer to natural conditions in groundwater and used magnesium sulfate solutions to compare the performance of cementitious binders in mortars.

9.2.2.2   Sulfate-resisting Portland cement.   The hydration products of a Portland cement with a low C_3A content have long been believed to resist damaging expansive reactions when in contact with sulfate-containing waters, and an ASTM limit of 5% (Bogue) was introduced in 1940. In the UK, a sulfate-resisting Portland cement is produced by reducing the alumina/iron oxide ratio of a raw materials mix to around 0.7. This reduces the potential C_3A content to a low level and reduces the Al/Fe ratio of the ferrite phase. Composition of this phase in sulfate-resisting clinkers is variable (Taylor, 1997), typically around $Ca_2 Al_{0.8} Fe_{0.8} Mg_{0.2} Si_{0.2} O_5$. The reduction in Al/Fe ratio is achieved by adding a by-product iron oxide to the kiln feed and, to keep the calcium silicates content of the clinker at a high enough level, a silica-rich material is incorporated (Section 2.1.3). This also effectively reduces the potential C_3A content by replacing some of the clay or shale.

An examination of the microstructure of hydrated pastes of a sulfate-resisting Portland cement (Gollop and Taylor, 1994) revealed that the ettringite formed in early hydration persisted for at least a year. Only minor amounts of monosulfate were detected. Although the sequence of chemical and physical changes, when pastes were immersed in sodium and magnesium sulfate solutions, were similar to those previously observed for ordinary Portland cement, they were less marked and accompanied by less decalcification of the gel matrix. The conclusion drawn was that these differences accounted for the greater resistance of hardened sulfate-resisting cement paste to attack, especially in young concrete where high porosity makes it particularly vulnerable (Gollop and Taylor, 1995).

The current British standard BS 4027: 1996 specifies a C_3A content of not more than 3.5% (on cement) and, because of this low level, a maximum of 2.5% SO_3 is allowed to regulate set. A number of blended cements containing ordinary Portland cement also provide sulfate resistance in mortars and concrete but specification of acceptable compositions is much more complex than for SRPC (Section 9.6). A European specification for sulfate-resisting cements, to be included in the EN 197 series, is still in draft form as inter-laboratory trials to select tests of sulfate resistance which will form a part of EN 196 are still in progress.

9.3   Cements with additional mineral constituents

The Type II designation of ENV 197 covers cements in which a significant fraction of the Portland clinker of a Type I cement is replaced by a second main constituent. Such cements are often referred to as *composite, blended* or *extended cements* but in ENV 197 the first term is reserved for cements containing more than one additional mineral. Type II covers a wide range of compositions, encompassing cements successfully

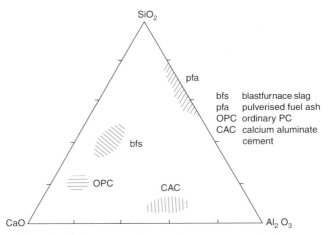

Fig. 9.2.   Mineral admixture and cement compositions in the ternary C–S–A system.   Sometimes CaO + MgO and Al_2O_3 + Fe_2O_3 contents are plotted in place of those of CaO and Al_2O_3

employed in Europe. With the exception of ground limestone, all the additional minerals contribute to concrete strength by chemical reaction and are low cost natural materials or industrial by-products. With the exception of silica fume (Section 9.4.4) compositions are bracketed in the same two broad ranges, A and B in Table 9.1.

Granulated blastfurnace slag is a hydraulic cementitious binder in its own right but requires an alkaline medium to initiate its hydration. The relationship of its composition of those of siliceous pulverised coal ash (fly ash) and Portland cement is indicated schematically in Fig. 9.2. It can be used at very high replacement levels with Portland clinker and these are covered by the Type III compositional specifications of ENV 197. Type IV — pozzolanic cements — contain mixtures of natural and other pozzolanic materials as other main constituents and Type V — composite cements — mixtures of blastfurnace slag and a natural or industrial pozzolana, or siliceous fly ash. The last two are not widely encountered and are not considered here.

9.4   Pozzolanic materials

In this section the origin, composition, hydration chemistry and practical value of pozzolanic mineral additions to Portland clinker will be briefly discussed. Dhir (1994) reviewed the physical and chemical requirements that ENV 197 places on these materials as well as their relevance in the UK, where historically they have usually been incorporated at the concrete mixing stage rather than in preblended cements. An appreciable difference in the level of available sulfate in a hydrating paste may exist between these modes of application, since only a preblended or interground cement will have a controlled SO_3 content. This difference may have significant practical effects on some concrete properties such as workability/water demand, early strength development and sulfate resistance.

9.4.1 Nature of the pozzolanic reaction

The basis of the use of pozzolanic materials in partial substitution for clinker in Portland cement is the ability of the alkaline medium produced by the hydrating cement to break down the silica or the alumina-silica networks in the particles they contain to form C–S–H. Reaction must occur at a practically useful rate and the reactivity of a pozzolan is primarily determined by its specific surface area and the proportion and reactivity of the glass it contains.

Crystalline quartz is made up of a fully condensed network of SiO_4 tetrahedra in which each oxygen is shared by (i.e. links) two silicon atoms; X-ray diffraction reveals the long-range order. In the reactive particles of a glassy pozzolana, the atomic networks lack the long-range order characteristic of a crystal. When a siliceous material has experienced a temperature high enough to induce partial or complete melting followed by rapid cooling, a high proportion of the particles are of a silica or alumina-silica glass (e.g. volcanic ash and pulverised fuel ash). In the case of clays and shales decomposed at lower temperatures, the reactive material is a powder consisting of very small, nearly amorphous particles (e.g. metakaolin formed in the decomposition of kaolinite).

The basic pozzolanic reaction involves the severing of Si–O–Si and Si–O–Al bonds by hydroxyl ions which can be schematically represented by:

$$\text{Si–O–Si} + 2\text{OH}^- \longrightarrow \text{Si–O}^- + {}^-\text{O–Si} + \text{H}_2\text{O} \qquad (9.2)$$

This is in fact the reverse of the dehydration accompanying the polymerisation process observed in a silica gel at a low pH but at that existing in cement hydration the surface hydroxyl groups are deprotonated (point of zero charge ca. pH 2 for surface Si–OH groups and ca. pH 9 for surface Al–OH groups). The possibility that a negative ion approaches a negative surface suggests a high activation energy for the reaction. However, at the concentrations of hydroxyl and positive counter ions in the cement pore solution, ion pair formation (e.g. KOH, $CaOH^+$) would presumably provide a more subtle mechanism of surface attachment.

Counterions for the negative charges left on the surface of a particle are provided by the cations in the pore solution of the hydrating cement namely: K^+, Na^+ and Ca^{2+}. The high surface area of the pozzolana favours

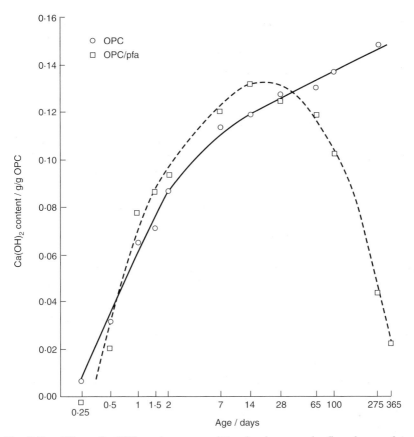

Fig. 9.3.   Effect of a 20% replacement of Portland cement by fly ash on calcium hydroxide formation in a paste (water/binder ratio 0.45) as a function of curing (Dalziel and Gutteridge, 1986)

the formation of non-swelling C–S–H with adsorbed alkali ions. Continued fracture of bonds in the surface of the solid will lead to the separation of silicate and aluminate ions from it. In the presence of an excess of calcium ions, derived from the calcium hydroxide produced by the hydrating cement, C–S–H and, by interaction with dissolved gypsum, calcium sulfoaluminate hydrates will be precipitated. Calcium hydroxide dissolution maintains the supply of the hydroxyl ions and calcium ions necessary; the overall reaction can be followed at a later stage in the hydration of the cement-pozzolana combination by the consumption of the crystals of this phase formed in the earlier and more rapid hydration of the Portland cement (Fig. 9.3). The role of the alkali ions is to increase the pH from that for saturated calcium hydroxide alone (ca. 12.5) to above 13.

In summary, to provide a useful level of pozzolanicity a material should possess an acceptable chemical composition, a high content of glassy/disordered phase(s), and a high specific surface, although not so high that it increases water demand unduly. To exploit the material a cement is required which produces a high pH in the pore solution of the paste and maintains an adequate supply of calcium ions. Clearly, for optimum performance, an efficient interdispersion of mineral addition and Portland cement is necessary, while curing conditions must ensure that premature drying of the paste is prevented. This is especially important since there is a slower initial development of strength than with plain Portland cement.

9.4.2   Fly ash (pulverised fuel ash)

9.4.2.1   Composition.   The exhaust gases from power stations burning pulverised coal carry much of the ash produced out of the boilers as a fine dust which is removed by electrostatic precipitators. The major part of the precipitated dust consists of glassy material derived from the clay minerals present in the coal employed and, unless combustion has taken place at an unusually low temperature so that little melting has occurred, most of the particles are spherical, a significant fraction being single hollow spheres (cenospheres) or hollow spheres with smaller spheres inside them (plerospheres). Particle sizes depend on those of the ground coal employed, the nature of the furnace and of the collecting equipment employed. Most lie in the range <1 to 200 μm with a maximum in the distribution around 10–20 μm.

In addition to clay minerals, the common impurities in coal are carbonates of calcium and magnesium (limestones and dolomitic limestones), sulfides (especially of iron), chlorides and quartz. Although the furnace temperature may exceed 1600°, much of the quartz is unchanged in the short residence time. Conditions in the flame are such that little exchange of material between particles is possible and considerable variation in composition both within and between individual glassy and crystalline particles is observed (Hemmings and Berry, 1988).

Thermal decomposition reactions of carbonates and clay minerals yield carbon dioxide and water vapour, respectively. Sulfides of iron are oxidised to hematite and magnetite (a spinel, Fe_3O_4) and sulfur dioxide. The production of these gases with the simultaneous formation of a viscous alumina-silica melt from the clay minerals provides the ideal conditions for *bloating*, i.e. thermal expansion of trapped gas. Hollow, low density particles, some of which are light enough to float on water, are formed on solidification. Cooling a melted clay mineral with either a two-layer (kaolinite) or a three-layer (illite) structure should yield crystalline silica and mullite as these are the stable sub-solidus phases in this part of the SiO_2/Al_2O_3 system. However, crystallisation of such liquids is slow, in part at least because of their high viscosity, and glass formation is favoured. Bulk analyses of UK ashes may be similar to those of dehydrated fireclays.

The use of fly ash as a partial replacement for Portland clinker in a blended cement can be seen as providing significant environmental benefits through a saving of the energy which would have been employed in producing and grinding the clinker substituted and a reduction in the amount of ash which must go to landfill. The ash may be either blended with Portland cement or interground with clinker and gypsum in a mill. Two types of ash are commonly available worldwide: from bituminous coals, a siliceous, low lime ash and from sub-bituminous coals and lignites, a calcareous ash. UK sources of coal provide only the former but imported sub-bituminous coals could yield calcareous ash. The ASTM designation for these two types is frequently encountered in the technical literature, namely: Class F for siliceous and Class C for calcareous ash, although these materials are actually defined as having silica + alumina + iron oxide contents above and below 70%, respectively. However, the reason for the reduced content of these oxides usually lies in the calcite content of the coal employed.

The European standard for fly ash, EN 450, specifies a content of 'reactive silica' of not less than 25%, less than 5% reactive lime in siliceous ash and a loss on ignition of not more than 5% (7% in BS 3892). In a siliceous ash this loss arises principally from the carbon content which, if too high, may lead to the discoloration of concrete or, by adsorption, reduce the effectiveness of any admixture (Section 10.1) employed in the mix. Methods of determining reactive (alkali soluble) silica and alumina and reactive calcium in a calcareous ash are given in the standard. The latter may have intrinsic hydraulicity as well as pozzolanic properties and for lime contents above 15% there is a 28-day strength requirement of $10\,N/mm^2$ when a ground sample is tested in a mortar according to EN 196. In order to avoid fluctuation in the quality of material received, especially in its carbon content, ash most likely to meet standards is that from a base load power station which employs one source of coal and maintains the continuous steady operation favouring efficient combustion. The 12.5% residue on a $45\,\mu m$ sieve, specified in BS 3892 Part 1: 1997, was originally introduced to monitor the fineness of an ash intended for incorporation at the concrete mixing stage, but the full particle size distribution or specific surface are more reliable guides to overall reactivity and effect on workability. EN 450 allows up to 40% residue on a $45\,\mu m$ sieve.

Cabrera et al. (1986) analysed several samples of fly ash from each of 18 UK sources and a summary of their results for the principal and usual

Table 9.3.   Oxide compositions of several pfa samples from each of 18 British power stations (after Cabrera et al., 1986)

Oxide	Overall mean: %	Range of means for 18 sources: %
SiO_2	51.4	47.9–54.9
SiO_2 alk. sol.*	29.9	23.3–33.7
Al_2O_3	27.0	22.1–28.9
Al_2O_3 alk. sol.*	5.49	3.79–7.83
Fe_2O_3	10.6	7.73–13.2
Na_2O	1.12	0.77–1.66
K_2O	3.33	2.19–3.81
CaO	2.40	1.22–6.42
MgO	1.62	1.39–2.31
SO_3	1.01	0.37–1.65
LOI[†]	4.43	2.13–12.73
Silica ratio S/(A + F)	1.37	—

* Soluble in alkali (BS 3892: 1988). This is taken to indicate the reactivity of the glassy phase.
[†] LOI: loss on ignition — principally carbon and adsorbed organic matter + water and carbon dioxide.

minor oxide components is given in Table 9.3. The observed standard deviations for oxide analyses for a single source were low, but for loss on ignition and for alkali solubilities of silica and alumina, standard deviations were significantly higher. The latter served as a measure of pozzolanicity and they noted that the UK standard applying at that time would have prohibited the use of two-thirds of their samples, although they yielded satisfactory concrete in performance tests. Almost 74% of their samples had a 45 μm residue in excess of the 12.5% specified in BS 3892.

Some 80–90% of the mass of a good quality *siliceous fly ash* consists of an alumina-silica glass, examination of which by ^{29}Si NMR revealed chemical shifts characteristic of a highly condensed network of MO_4 tetrahedra, with actual values influenced by substitution of silicon (4+) by some of the aluminium or iron (3+) in some tetrahedra (Bijen and Pietersen, 1994). Such a substitution leaves a tetrahedron with a net negative charge, so that an oxide ion bridge to an adjacent tetrahedron is broken and a charge compensating cation incorporated. The cations commonly available in fly ash, other than a proportion of the aluminium, are Ca^{2+}, Mg^{2+}, Na^+ and K^+ derived from the impurities in the coal. In the glass they are associated with the non-bridging oxide ions and have higher coordination numbers than the network-forming elements (6 or higher). Small amounts of crystalline components are also found, either as inclusions in larger, partially devitrified particles attached to the glass spheres, or as discrete angular particles. Phases detected by XRD include:

quartz; mullite (A_3S_2); the iron oxides, hematite and magnetite; anhydrite and potassium sulfate. Apart from the sulfates, these phases are merely inert diluents. The glass superimposes a broad halo on the XRD pattern with a maximum which ranges from 25 to 30° 2θ with increasing calcium content.

A *calcareous fly ash* may contain more than 20% lime and the composition lie in the gehlenite (C_2AS) primary phase field. The proportion of crystalline phases is greater than in a siliceous ash and may include periclase, anhydrite, ferrite spinel and merwinite (C_3MS_2) but the major component is a reactive, calcareous, glassy phase. At the highest levels of calcium, free lime and C_3A may be present in an ash and cause storage and potential unsoundness problems.

9.4.2.2 Hydration. This account will be limited to siliceous (low calcium) ashes. Except where it contains a high carbon content or a high proportion of very coarse particles, substitution of fly ash for cement in a concrete or mortar generally results in a reduction in water demand and an increase in the cohesiveness of the fresh mix up to a limit at about 20% ash content. Reduced water demand and improved flow have usually been ascribed to the effect of the spherical shape of the ash particles. However, from an examination of published data, Helmuth (1986) concluded that the effect was the result of adsorption of submicron, negatively charged particles of the silica-alumina glass on the surface of the cement grains. Mortar or concrete flow is thus improved by the mutual repulsion of the negatively charged particles in the paste fraction, a mechanism analogous to that provided by some plasticising admixtures (Section 10.1). If advantage of the effect is taken in concrete mix design it can partly compensate for the reduction in early strength which results from the substitution of clinker by ash.

Initial setting times of Portland fly ash cement pastes are longer than for equivalent plain cement pastes, although this effect is smaller than those of cement fineness, temperature and water/binder ratio. The rate of dissolution and solubility of an ash in the aqueous phase of a hydrating cement is strongly dependent on its pH and only increases significantly above a value of about 13 (Bijen and Pietersen, 1994). For the samples they examined, differences in chemical composition were a far less significant influence on ash reactivity than their glass content and specific surface area.

After the fall in pH at the end of the dormant (induction) period in the hydration of a cement (Section 7.3), it rises again as leaching of K^+ and Na^+ ions from the clinker phases continues and the volume of pore solution diminishes, as water is increasingly incorporated in solid hydration products. Depending on temperature and water/binder ratio, an accelerating rise in pore solution pH is observed in the period 1–3 days. This second induction period is reflected in the percentage of ash consumed as a function of time. Initially, the formation of solid calcium hydroxide from the hydration of the alite in the cement far exceeds its consumption in the pozzolanic reaction, but eventually the amount in the hardened paste peaks and then declines. Fig. 9.3 shows this effect for a particular paste composition examined by Dalziel and Gutteridge (1986) for which they found a consumption of 0.7 g of portlandite per g of fly ash reacted after 1 year of hydration. QXDA revealed that the early hydration of alite was accelerated somewhat in the presence of the ash, while after 14 days that of belite was strongly retarded.

The principal hydration product in a Portland cement/fly ash paste is C–S–H. Its composition is variable but from an examination of pastes by SEM microanalysis, a mean C/S ratio of around 1.6 has been reported at the 20% fly ash level (Harrisson et al., 1986). At higher levels of clinker replacement and at long periods of hydration, lower C/S ratios have been observed. The major aluminous hydration product detected in the early stages of hydration is ettringite (AFt) and at later ages an AFm phase. The balance between the total reactive alumina and iron and the total sulfate (including that available in the ash) should determine the final outcome of the hydration reactions, but how far equilibrium is approached in these heterogeneous systems depends on the level of interparticle dispersion and the initial water/binder ratio. Other hydrated aluminates such as strätlingite (C_2ASH_8) and a hydrogarnet have been detected by XRD in older pastes. Long term results reported by Lachowski et al. (1997) for the hydration of fly ash/cement pastes have indicated that equilibrium may not be reached even after 6 years' curing at 98–100% relative humidity (Fig. 9.4).

The principal heat evolution peak (II in Fig. 7.1) is delayed by 3–4 hours in the hydration of a 30% fly ash/cement blend and the maximum depressed, but this early retardation is followed by an acceleration of the hydration of alite and heat liberation rate is similar to that of plain Portland cement after 1 day. The rate of dissolution of an ash in the pore solution of cement paste is markedly accelerated if the temperature increases above 40°, a temperature easily exceeded in a large concrete pour. Consequently, if a Portland fly ash cement is to be used as a low heat cement to control overall temperature rise in a large structure, an ash content of at least 75% is likely to be required (Kaushal et al., 1989). Unreacted ash in such a concrete is regarded as an acceptable microaggregate.

Fly ash begins to make a contribution to the compressive strength of a mortar or concrete after 7–14 days at ordinary curing temperatures, depending on its fineness and glass content and the reactivity (C_3S

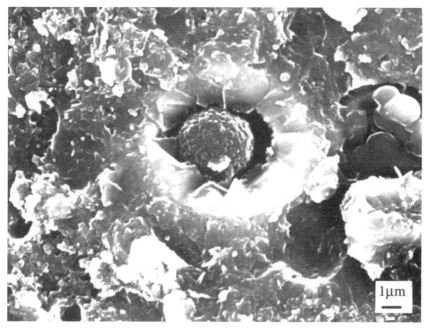

Fig. 9.4. Scanning electron micrograph showing an unreacted fly ash particle in a blended cement paste hydrated for six years (courtesy of E.E. Lachowski and K. Luke)

content) of the cement. Early strength is particularly dependent on the sulfate content of the mix. Dalziel and Gutteridge (1986) compared the growth in compressive strength of 0.47 water/binder ratio pastes, one of ordinary Portland cement, the other with a 30% substitution of cement by fly ash. After a year the latter had not quite caught up with the pure cement paste (63.2 compared to 65.4 N/mm^2). In mortars and concretes, on the other hand, it has been found that similar replacement levels result in catching up and overtaking in 3–6 months. Partial substitution of fine and even coarse aggregate by ash has a beneficial effect at even earlier ages (Jawed and Skalny, 1991). The greater benefit observed in adding ash to mortar or concrete as compared to a cement paste is attributed to the reduction it produces in the porosity and portlandite content of the transition zone around the aggregate (Section 8.1).

The replacement of calcium hydroxide by C–S–H in a hardened paste provides the potential for producing a low permeability mortar or concrete by modifying the pore size distribution. The fine gel porosity of C–S–H will replace some of the coarser porosity of clusters of portlandite crystals. The effect is revealed by the change in the distribution of fine and coarse pore volumes in Fig. 9.5. One result of this is that chloride ion

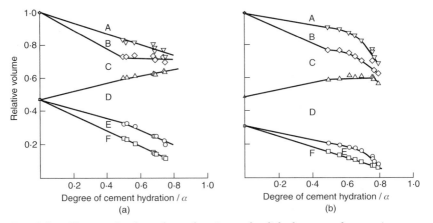

Fig. 9.5.   Changes in the volume fractions of solid phases and pores in pastes with increasing degree of hydration at a water/binder ratio of 0.35: (a) Portland cement; (b) same cement blended with 30% fly ash. The effect of the ash at higher degrees of hydration is a replacement of coarser pores (>4 nm) by finer, gel pores (Patel et al., 1989). Porosity determination by methanol adsorption (Section 8.2.3): A — pores < 4 nm; B — pores 4–36 nm; C — pores > 4 nm; D — gel + ash; E — calcium hydroxide; F — anhydrous cement

diffusivity is strongly reduced by incorporation of fly ash in a hardened paste (Page *et al.*, 1981, Malek *et al.*, 1989). However, a reduction is observed even when a paste containing ash has a higher porosity than a reference plain cement paste and it is ascribed to the effect of a specific interaction of the chloride ion with alumina-containing phases, the contents of which are increased by the introduction of the ash. The electrical conductance of a paste is principally determined by the mobility of ions in the aqueous pore solution and it provides an indication of pore continuity which is more directly relevant than pore size distribution. The higher electrical resistance of a concrete containing 25% fly ash after three weeks' hydration, compared to that of a reference concrete, indicated more effective pore blocking resulting from the microstructural modification produced by the ash (Fraay *et al.*, 1989).

9.4.3   Natural and related pozzolanas
Natural pozzolanas are of three main types, as detailed below.

(*a*) Those of the first type, described by Sersale as true pozzolanas, are powders deposited by an explosive eruption of a volcano. They are classified on the basis of particle size (Tucker, 1981) and the smallest particles may remain suspended in the atmosphere for long periods, travelling considerable distances. When water is available in a lake nearby or heavy rainfall accompanies an eruption, ash particles may be transported as a mud. The rapid cooling which explosively ejected magma experiences means that the smaller particles are mainly composed of a reactive glass, with only a small content of minerals such as quartz or a feldspar. Non-explosive eruptions yield coarser less reactive materials because cooling is slow and/or the chemical composition does not favour glass formation.

Composition ranges (Sersale, 1980) reveal the relationship to dehydrated clay minerals:

SiO_2 45–60%; Al_2O_3 + Fe_2O_3 15–30%; CaO + MgO + alkalis ca. 15%; LOI <10%

Particles are porous as a result of gas exsolution from the depressurised magma and, if this is viscous, highly porous pumice is produced. The glass is highly reactive, containing silica and alumina which are readily soluble in alkali. Natural weathering results in the formation of zeolites which contain three-dimensional aluminosilicate anions with exchangeable cations such as K^+ and Ca^{2+} in pores, the shapes and sizes of which are defined by the crystal structure. The siliceous zeolites are very reactive and contribute to the overall pozzolanic behaviour.

(b) Weathered, consolidated sediments of ash (<2 mm) known as tuffs may contain sufficient reactive glass and zeolites to be valuable pozzolanas. An example of this type in the lower Rhine basin is Rhenish trass.

(c) The third type of natural pozzolana is constituted of mainly siliceous matter in the form of sediments of skeletons of life forms such as diatoms. These may have silica contents of over 90% and are described as earths, loams or even (misleadingly) clays. While they react readily with lime, their fineness and form has a deleterious effect on water demand which restricts their use. A silica-rich material produced by ashing rice husks is also used in those parts of the world where rice is a major crop.

Clay minerals, the ultimate weathering products of igneous rocks, react with lime to form hydration products similar to those formed from Portland cement, but inhibit its normal setting and hardening. Lime and cement are, however, valuable in soil stabilisation. If clays or shales are subjected to moderate heat treatment to break down their crystal structure by dehydration, useful pozzolanas are obtained. Burnt oil shales provide a by-product source and such materials are included in ENV 197. The use of all these materials in concrete can have a significant economic advantage if this is not offset by: the costs of extraction or recovery of a by-product; treatment, including grinding if necessary; and transport to the point of use.

In general, natural pozzolanas are more reactive than pfa, yielding similar hydration products: C–S–H, and calcium aluminate hydrates, sulfoaluminate hydrates and possibly also strätlingite or hydrogarnets. The similarity extends to inhibition of βC_2S hydration. Their use in concrete has included limiting temperature rise and reducing alkali-silica expansion. Recommendations and guidance for their use are given by the American Concrete Institute (1994).

9.4.4   Silica fume (microsilica)

In the production of silicon and ferrosilicon alloys by reduction of silica with carbon in an electric furnace, some SiO vapour is formed. This by-product oxidises in the vapour phase on contact with air and the resulting fume is condensed to yield a very fine silica powder, typically with a particle size of 100–500 nm and a nitrogen BET surface area around 20 000 m²/kg. Quality varies with source, the best material having silica contents in the range 90–94%, although Wolsiefer *et al.* (1995) made concretes giving similar performance test results with silica contents in the range 79–95%. However, even a surface coating of carbon can greatly reduce pozzolanic activity. The term microsilica is widely used and includes very pure silicas such as those manufactured as reinforcing fillers for plastics. These can have surface areas above 150 000 m²/kg.

Condensed silica fume (CSF, an abbreviation widely used in the literature) has a very low bulk density with values down to one-tenth of the true particle density of silica glass, 2300–2400 kg/m³. It is frequently transported in the form of compacted granules, as pellets, or as an aqueous slurry with an added dispersant such as a superplasticiser (Section 10.1). Minor impurities include: alumina, magnesium oxide, lime and alkalis and possibly some iron and iron silicide. Pelletised silica fume is suitable for milling with Portland clinker in a cement mill. Granules may be broken down and dispersed in a concrete mixer by the aggregate. Aged agglomerated aqueous suspensions may need to be redispersed before use. Wolsiefer *et al.* used sixteen samples from ten North American sources covering this range of possibilities. Portland/silica fume cement is specified in ENV 197 as a blend of 6–10% with clinker (Table 9.1). The silica content of the fume must exceed 85%, the specific surface exceed 15 000 m²/kg and the loss on ignition be less than 4%.

Silica fume (SF) particles possess no long range order and ²⁹Si NMR reveals that it is highly condensed with bulk silicon atoms linked to four neighbours by bridging oxygen atoms (Bijen and Pietersen, 1994). After prolonged soaking in water, surface oxygens in such a silica are present as silanol (Si–OH) groups. It is highly reactive and, in a blended Portland cement paste, the portlandite content may pass through a maximum in under one day as it is consumed in the pozzolanic reaction which, if the SF is well dispersed, is likely to be largely completed after two weeks. Rates of reaction of individual phases in a silica fume cement paste can be followed by a combination of ^{29}Si NMR and thermal analysis (Justnes et al., 1993). Taylor (1990) calculated that up to about 35 g of microsilica can react with 100 g of Portland cement, assuming that the C/S ratio of the C–S–H formed cannot fall below 0.8.

The use of silica fume in concrete is limited by its cost and availability. To offset the increased water demand introduced by this fine material, a superplasticiser is incorporated into a concrete mix and the accelerating effect of silica on alite hydration may make the inclusion of a retarder desirable (Section 10.1). The increase in cohesiveness of a mix it produces can be of value, in underwater concreting for example (Neville, 1995). Efficient dispersion of silica fume in a concrete is not only necessary to ensure maximum benefit from its contribution to properties such as strength and permeability but also to avoid a possible alkali-silica reaction in which an agglomerate of silica particles is big enough to act as an especially reactive aggregate forming an alkali-rich swelling gel (Lagerblad and Utkin, 1994).

The high reactivity of silica fume leads to an increase in early heat liberation in a concrete, although as a replacement for Portland cement SF progressively reduces total heat liberation as levels increase above 10%. Incorporation of silica fume can enhance concrete strength (increases by a factor of around 2 are possible) but the principal practical benefit is reduced permeability and hence improved durability. The latter derives from the acceleration in the early hydration of alite combined with the high pozzolanic reactivity of the silica which accelerates the consumption of portlandite and the formation of C–S–H. Total porosity does not appear to be reduced significantly by the pozzolanic reaction but, as observed with fly ash, the pore size distribution is modified by an increase in the proportion of fine pores in the C–S–H gel. Some larger but unconnected pores are formed where portlandite crystals have dissolved and reacted with the silica. The densification of the transition zone at the paste–aggregate interface can contribute to a reduction in the permeability of a concrete by a factor of around 100.

9.5   Blastfurnace slag and blastfurnace slag cements

9.5.1   Composition

In the production of iron in a blast furnace, a flux such as limestone or dolomite (and sometimes bauxite) is added to the charge of iron ore and coke. This produces a liquid slag at 1400–1550° which contains the siliceous and aluminosilicate impurities originating in the iron ore and the

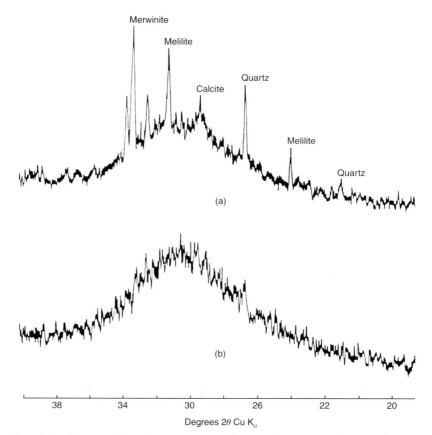

Fig. 9.6. X-ray diffraction patterns of two slags: (a) glass with some crystallisation; (b) glass halo (Moranville–Regourd, 1998)

coke and which is immiscible with the liquid iron. The principal oxide components of the slag are thus lime, silica and alumina and compositions are often represented in a ternary diagram such as Fig. 9.2, although significant amounts of magnesia (up to ca. 20%) may be present when dolomite is used. Slow cooling of the slag after it is tapped from the furnace allows crystallisation to occur and the product has little or no cementitious value. The principal crystalline phases usually formed are merwinite (C_3MS_2) and melilite, which is a solid solution of gehlenite (C_2AS) and åkermanite (C_2MS_2). However, rapid cooling of the liquid slag, involving mechanically converting it to small droplets (granulation) or pellets, inhibits crystallisation (Fig. 9.6). Granulation produces a particularly high proportion of glass (usually >90%) and, after grinding, a product which resembles sand and has latent hydraulicity is obtained (ground granulated blastfurnace slag: ggbs). As the term 'latent' implies,

Table 9.4.   Composition ranges for 16 granulated blastfurnace slag sources in seven European countries (Livesey, 1993)

Oxide	Range: %	Oxide	Range: %
SiO_2	33.9–38.1	K_2O	0.31–0.72
CaO	36.6–42.8	Na_2O	0.20–0.45
Al_2O_3	8.8–13.3	SO_3	0.0–2.4[†]
MgO	6.7–12.8	S^{2-}	0.8–1.3
Glass content	89–99*		
Ratios			
(C + M)/S	1.27–1.47		
(C + M + A)/S	1.53–1.85		

* 15 samples — one sample had a glass content of only 75%.
[†] 15 samples: $9 \leq 0.1\%$; $6 > 0.5\%$.

an activator is necessary for this property to be developed and an alkali (Section 10.4), lime, gypsum and Portland cement are effective, although major usage involves only the latter. In its application, ggbs may be blended with Portland cement or interground with clinker to an optimum SO_3 level.

Slag cements have been produced since before 1900 in continental Europe and ENV 197 covers the wide range of compositions commercially available (Table 9.1). In the UK, the great majority of ggbs is used by direct introduction into concrete in the 'ready mixed' industry. The required chemical and physical properties of the slag for this application are specified in BS 6699: 1992. The composition of slag from a particular plant depends on the purity (source) of the iron ore (usually an iron oxide — hematite or magnetite), the coke and flux employed and, unless sources of these are changed, variability is limited. However, compositions vary from one plant to another and the range encountered worldwide is considerable. Data for some European plants are given in Table 9.4. The reducing conditions in the blastfurnace, produced by the carbon monoxide formed by the partial combustion of the coke in the charge, result in any iron and manganese in the resulting glass being in the divalent state and some metallic iron may also be present, although usually these are limited to tenths of one percent and slag is almost white in colour. The magnesium oxide content (limited to 14% in BS 6699) does not cause a soundness problem because the magnesium ion is distributed in the glass and not present as the mineral periclase (Section 6.6). Much of the sulfur is present as the reduced form S^{2-} and this is contained in the glass. Hydration of the slag in the presence of cement produces a blue-green colour which subsequently fades as oxidation takes place.

ENV 197 specifies that a blastfurnace slag for blended cement manufacture must have a glass content of at least 66.6% and that the sum of the oxide mass percentage contents ($CaO + MgO + SiO_2$) be at least 66.6%. In addition, the ratio ($CaO + MgO$)/SiO_2 must exceed 1.0. The requirement for the remainder is simply that it be Al_2O_3 with only small amounts of other oxides. A number of attempts, with limited success, have been made in the past to correlate the oxide composition of slags with their potential hydraulicity (Smolczyk, 1980). The basicity ratio or modulus, usually written as ($C + M + A$)/S, where the symbols represent the mass percentages of these oxides in the slag, has been the most commonly employed and hydraulicity rated as good, moderate or poor when the value lies above 1.0, from 0.5 to 1.0, and below 0.5, respectively. Minimum values of 1.0 or 1.4 have been specified in some national standards.

The constitution of the glass in blastfurnace slag has been examined by TMS (Section 7.2) and Raman and NMR spectroscopy. These techniques have shown that it contains isolated SiO_4^{4-} and dimeric $Si_2O_7^{6-}$ anions and Ca^{2+}, Mg^{2+}, Al^{3+} and $Al-O^+$ as cations. A review of the structural aspects of this glass (cation oxygen coordination numbers, microheterogeneities, and structural and surface defects) and the physical methods employed in their examination has been given by Regourd (1998). Crystalline material, when present, is identified by optical microscopy and X-ray diffraction (Fig. 9.6). Merwinite and melilite are often found as inclusions in the glass, in sizes ranging from dendrites less than 100 nm wide to crystals visible in the optical microscope. Metallic iron may also be present as inclusions.

9.5.2   Hydration

The reactivity of a slag depends on its glass content and the nature and concentration of the defects present in this phase, both of which are affected by the rate of cooling to below about 800° which it has experienced. As in all phase-interface reactions, the particle size distribution (fineness) of a slag is also a primary variable. Spectroscopic techniques such as X-ray fluorescence have been used to establish that magnesium and aluminium ions in slags and synthetic glasses in the C–M–A–S system are located in both four coordinated (tetrahedral) and six coordinated (octahedral) sites and hydraulicity has been linked to the content of the latter. Regourd (1986) considered that the hydraulic reactivity of a slag is determined by the proportion of octahedral aluminium in the glass which dissolves first in an alkaline medium, the calcium silicate network being hydrolysed subsequently. A pH above that which calcium hydroxide can provide is not required to initiate slag reaction as it is for fly ash. Together with the role of structural defects, including incipient crystallisation, the several factors influencing slag hydration make it hardly surprising that simple bulk composition-reactivity relationships have proved of limited value.

The early hydration of Portland cement clinker is accelerated in the presence of blastfurnace slag and the rate of hydration of the latter is dependent on the efficiency of its dispersion (Lee *et al.*, 1987). Lumley *et al.* (1996) used a solvent extraction method (Section 4.2) to determine the unreacted slag in pastes of several slag cements hydrated at 20° at water/binder ratios of 0.4 and 0.6. They found that the percentage of the slag hydrated at 28 days was 30–55% and at 2 years 45–75%. These values are similar to that for belite (βC_2S) in Portland cement. The slags they examined differed significantly in reaction rate which, for their samples, was related to differences in hydraulic moduli and fineness. Reaction rate of a slag decreased with decreasing water/binder ratio and with an increasing proportion of slag in the blended cement.

The principal products of hydration of a slag cement are C–S–H (C/S ratios in the range 0.9–1.6 have been found), AFt and/or AFm phases, but with the portlandite expected from the Portland cement component reduced in quantity by interaction with the slag. Microanalytical determination of Al/Ca ratios in the electron microscope suggested that C–S–H and AFm phases are intermixed on a submicron scale (Harrisson *et al.*, 1987). With high slag proportions, the amount of portlandite formed may pass through a maximum as hydration proceeds. Strätlingite (C_2ASH_8), siliceous hydrogarnets (solid solution series C_3AS_3–C_3AH_6) and a hydrotalcite type phase are formed. Hydrotalcite phases vary in composition and are related to brucite ($Mg(OH)_2$), having a layer structure in which some of the Mg^{2+} ions are replaced by Al^{3+} and some of the OH^- ions by CO_3^{2-}.

Hydration products form separately at the surface of both clinker and slag particles and extend inwards with time. The level of heterogeneity found in an examination of the microstructure depends on the mobility of the ions involved in an alkaline medium. Bijen and Pietersen (1994) found that a net movement of calcium, silicon and aluminium out of the slag occurred while magnesium acted as an immobile marker. This would be expected from the low solubility of the magnesium ion in an alkaline medium. They also observed that the reaction of the slag was initiated at the onset of cement hydration, at a pH around 12, and that the subsequent rise observed with a neat cement paste did not occur. Glasser *et al.* (1987) noted that pH rises little above 12 and that a slag/cement paste also differs from that of neat Portland cement paste in producing a chemically reducing environment in the pore solution, involving the S^{2-} ion, when the slag content is as high as 70%. On hydration, the sulfide is considered to be located in the AFm phase and is eventually oxidised to sulfate. Lachowski *et al.* (1997) found the development of the microstructure in a slag/Portland cement blend to be continuing after six years' curing at 98–100% rh, with unreacted slag still present.

Because ggbs does not rapidly react with (or absorb) water, its substitution for Portland cement has a beneficial effect on the workability of fresh concrete. Since, however, grinding slag produces irregular glassy shards, particle shape makes no positive contribution to flow as has been suggested for fly ash. The slow, early reaction of a slag in a blended cement results both in a slower decline in workability (slump loss) and a slower heat build-up than in the equivalent plain Portland cement concrete. However, the hydration of a slag is significantly accelerated if the curing temperature is raised above ambient to around 40° and, since this is a temperature easily reached in a large concrete structure, an isothermal laboratory evaluation of the strength development and heat liberation rate in a concrete made with a slag cement will seriously underestimate performance in the field (Neville, 1995). Conversely, the growth in strength of a slag cement will be poorer than that of an ordinary Portland cement at low temperatures. The relationship between slag/ Portland cement ratio and ultimate temperature rise is such that at low to moderate slag contents (20–45%) the temperature rise in a $2\,m^3$ concrete pour may be greater than that in a control concrete (Coole, 1988). However, at high slag contents the short-term heat output decreases and blends with a very high slag content have been used, in countries such as Japan, as super low heat cements for the construction of dams.

In general, laboratory tests at ambient temperatures reveal a reduction in concrete and mortar strength at one day, deriving from the substitution of slag for Portland clinker but with an optimum gain from about seven days onwards, depending on slag reactivity and content. The role of the C_3S content of the Portland clinker in activating the slag in a blend is indicated by results obtained by Kelham and Moir (Fig. 9.7). The early strength developed by a slag cement has been shown to be increased significantly by increasing its SO_3 content from 0.75% to 2% (Frigione and Sersale, 1983) but this did not result in an optimum 28-day strength. With slags ground and then passed to a classifier to obtain a fine particle fraction, very high strength concretes can be produced. A dispersing agent is then necessary in a concrete formulation to aid workability at an acceptable water/binder ratio and to prevent agglomeration of slag particles, which will reduce their interaction with the pore solution. With proper aggregate–cement proportions and attention to curing, high strength concrete with an advantageous pore size distribution and a low permeability can be produced (Nakamura et al., 1992).

9.6   Problems of specification of blended cements

The use of blended cements has economic, environmental and in some applications distinct practical advantages. However, in spite of many

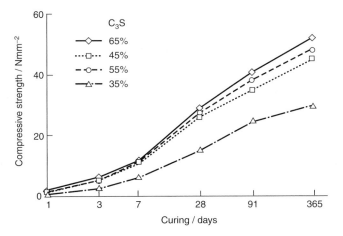

Fig. 9.7. *Influence of the C_3S content of Portland cements on concrete strength development in blends with 70% blastfurnace slag; water/binder ratio = 0.6. The cements were synthesised with a range of C_3S contents but all with approximately 75% total silicates (Kelham and Moir, 1993)*

investigations, two particular applications have raised as yet unresolved problems in specifying the compositions of constituents needed to provide an acceptable performance of their blend in a concrete. These relate to their contribution to sulfate resistance and to the inhibition of alkali-aggregate reaction in concrete. A positive contribution is to be expected from the effect of mineral additions on the pore size distribution of the paste fraction of a concrete where the extent of the transition zone is much reduced as coarse calcium hydroxide crystals are replaced by C–S–H. This reduces permeability and a denser surface layer of paste on a concrete further inhibits the ingress of aggressive agents.

9.6.1 Sulfate-resistance of blended cements

Laboratory assessments of the performance of blended cements exposed to sulfate solutions have shown that the substitution of some of the Portland cement by blastfurnace slag or fly ash may improve or worsen attack, depending on the composition of the materials and the test employed. For example, Locher (1966) found that slag substitution at the 65% level improved resistance to sodium sulfate solution and this was independent of the alumina content of the clinker employed in producing the blended cement. However, whether the resistance of a cement with a lower slag content was better or worse than the corresponding plain cement was dependent on the alumina contents of both clinker and slag.

Tikalsky and Carrasquillo (1993) assessed the performance of binders prepared with 25% fly ash substitution of Portland cement using a wide range of fly ash compositions. The expansion and cracking of concrete samples exposed to 10% sodium sulfate solution was found to be related to the lime and alumina contents of the glassy phase in the ash. Such results have indicated the importance of the reaction of aluminates with sulfate in deterioration. The effect of lime in a fly ash was considered by Tikalsky and Carrasquillo to derive from an enhancement of the reactivity of the glass component.

Lawrence (1992) examined cements containing fly ash, silica fume or slag and noted that they are all significantly more susceptible to a magnesium sulfate solution than to one of sodium sulfate. He compared blended cements with sulfate-resisting Portland cement in long-term exposure of mortars to sulfate solutions under laboratory conditions and demonstrated the superiority of the latter, although the low permeability and reduced portlandite content of *adequately cured* blended cements has been shown to resist penetration of chloride and weakly acidic solutions. Blends of sulfate-resisting cement with mineral admixtures do not appear to have consistently exhibited the expected advantages (Bensted, 1995).

The difficulty in formulating compositional requirements as part of a specification of sulfate-resisting blended cements stems from the number and complexity of the factors influencing their performance and the limited test regimes which have usually been employed (Lawrence, 1990). The factors influencing sulfate attack in these systems are: the proportion of the mineral component and its chemical and phase composition; the composition of the Portland cement used; the nature and variability of the hydration products (microchemistry and microstructure); the aggressive sulfate involved and its concentration in solution; the temperature of exposure; and the overall SO_3 content of the blended cement. Sensitive variables in test methods, such as the curing employed for test specimens, also introduce problems (Section 9.2).

Most of the investigations published have involved simply blending the mineral admixture with a Portland cement, although the value of targeting the SO_3 content has been demonstrated. Kollek and Lumley (1990) examined the long-term resistance of slag cements to both sodium and magnesium sulfates and noted the importance of intergrinding to an optimum SO_3 content. It was possible to correlate the hydraulic modulus of the slags they examined with the rate of hardening of their samples and, while strength development improved with an increasing alumina content of the slag, the reverse was true of sulfate resistance.

Lawrence pointed out that particularly comprehensive testing of blended cements is necessary to ensure their exposure to a range of conditions, under any of which a particular composition might fail. He concluded that testing should involve: a number of sulfate types and solution concentrations; a range of temperatures of exposure (including low temperatures); partial immersion of specimens; and both strength and expansion measurements. These conclusions were based on observations such as the much greater sensitivity of slag cements to magnesium sulfate than sodium sulfate solutions; the different effects of change of exposure temperature on the relative performance of sulfate-resisting Portland cement and blended cements containing different mineral constituents; and the fact that some slag cements soften without expansion.

Microchemical and microstructural changes in hardened slag cement pastes immersed in sodium and magnesium sulfate solutions have been examined by Gollop and Taylor (1996). Their results suggested that the origin of the complex relationship between the nature of the sulfate solution and the slag cement composition might lie in the fact that both influence the relative seriousness of two mechanisms of paste degradation. As with ordinary Portland cement, but with morphological differences, these are:

(a) an expansive conversion of an AFm phase to ettringite with some decalcification of the C–S–H gel matrix and
(b) decalcification of the matrix with the formation of gypsum. In the case of magnesium sulfate, which is particularly destructive of C–S–H, brucite and a hydrated magnesium silicate are also formed at heavily attacked edges of specimens.

Conflicting results have been reported on the value of silica fume in sulfate resistance. Additions of 10–25% silica fume to Portland cement have been shown to increase the resistance of mortars to sodium sulfate, but a strong magnesium sulfate solution attacked the same blends more vigorously than plain cement mortars, with considerable loss in the mass of exposed specimens (Cohen and Bentur, 1988). Lawrence (1992) found that a 10% silica fume blend performed satisfactorily over a period of two years in magnesium sulfate solution containing 24 000 mg/l sulfate ion.

Guidance on the specification of concrete to be exposed to sulfate-containing groundwater is given in the UK in BS 5328: Part 1: 1997. Account is taken of the uncertainties surrounding the performance of blended cements. Blended cements (25–40% pfa or 70–90% ggbs with ordinary (42.5 N) Portland cement) are considered acceptable alternatives to sulfate-resisting Portland cement at concentrations up to 6000 mg/l SO_4^{2-} and 1000 mg/l Mg^{2+} or 3000 mg/l SO_4^{2-} if Mg^{2+} exceeds 1000 mg/l. Above these levels, SRPC must be employed. In view of their importance in determining the permeability of concrete, minimum binder contents and maximum water/binder ratios are also recommended.

9.6.2   Influence of mineral additions in blended cements on the alkali-silica reaction

Both Portland cement and fly ash (or slag) contain alkali, much of which is slowly released as hydration of the paste of a blended cement proceeds. The concentration of the ions in the pore solution will also rise as an increasing proportion of the pore water is chemically bound in the hydration reactions. The origin of the problem of linking performance with composition lies in the complexity of the mechanisms determining the rate of release and the proportion of the total alkali present in the constituents of a blended cement entering the pore solution of a hardened paste. Apart from the rapidly dissolved alkali sulfates in the clinker, alkali release from solid solution in the clinker phases will occur as they hydrate, and from belite this will be particularly prolonged. The rate of release of alkali from the glassy phase of the ash or slag will depend on its composition and reactivity. Offsetting these additional sources of alkali will be a possibly greater retention of alkali ions by the low lime C–S–H produced in a pozzolanic reaction. A further complication follows from the fact that while alkali contents of the materials in a blended cement-aggregate system are determined to indicate the potential for alkali-silica reaction, the primary aggressive agent is the hydroxyl ion. As pointed out in Section 9.2, the alkali metal cations have a secondary role in increasing the pH above that for saturated calcium hydroxide but the magnitude of the effect is modified if anions other than OH^-, such as sulfate, are involved in the charge balance in the pore solution.

A direct evaluation of the potential performance of a blended cement involves examining mortar prism expansion when it is employed with a known reactive aggregate. Opaline silica has been used to achieve accelerated reactivity, but a number of workers have pointed out that this aggregate may be too reactive. Most natural aggregates induce expansion slowly and incorporation of opaline silica may result in an unrealistic ratio of the rates of the alkali-silica and the pozzolanic reactions (Thomas, 1994).

Hobbs (1994) reported expansion measurements for concrete specimens made as part of an ongoing British Cement Association programme, which reveal the interactive behaviour of the alkali sources in a blended cement. Concretes were made with: three samples of Portland cement (containing 0.6, 0.9 and 1.2% Na_2O equivalent); three samples of fly ash (containing 3.0, 3.3 and 3.9% Na_2O equivalent); and three levels of cement replacement (6, 25 and 40%) with Portland cement alone in the plain concrete control. The aggregates employed were Mountsorrel granite and BS 4550 sand with a partial replacement of the latter, in appropriate size ranges, by three levels of cristobalite, a high temperature polymorph of silica susceptible to expansive reaction. Results obtained after two years showed that at an Na_2O equivalent in a concrete of 3 kg/ m^3, the effectiveness of an ash in reducing expansion depended on the total alkali content of the ash, the proportion of Portland cement replaced and the alkali content of the cement. At the 40% replacement level, all the fly ash samples reduced expansion due to alkali-silica reaction with the cements employed. After two years, only the concretes made with the ash having the highest alkali content (3.9% Na_2O equivalent) showed unacceptably large expansions. Hobbs estimated that the effective alkali contribution from this ash was zero when used at this level of cement replacement with the 1.2% alkali Portland cement, 3% of its total alkali content when used with the 0.9% alkali Portland cement, and 9% of its total when combined with the 0.6% alkali Portland cement.

Lachowski *et al.* (1997) commented that while there is evidence that mineral additions such as fly ash and slag can inhibit or even suppress alkali attack on a siliceous aggregate, there is no agreement on their efficiency in doing so. They examined pore solutions expressed from plain and blended cement pastes after two and six years' curing at 98–100% rh. Two fly ash samples and one blastfurnace slag each blended with the same Portland cement were examined. Total alkali concentrations were reduced in the presence of fly ash and ggbs, but sodium ion concentration actually increased in two pastes. All the pore solutions had a pH of 13.1–13.6 and all the pastes contained residual calcium hydroxide but at a reduced level in the blended cement pastes. A considerable difference in the reactivity of the two fly ash samples was indicated by the continuing reduction in portlandite content of one of the pastes between two and six years. The authors also noted that the prolonged release of alkali from their sample of cement had not been observed with other samples they had examined, where concentrations had reached a steady state after three months.

From such observations it must be concluded that while uncertainties remain concerning mechanisms of alkali release from reacting solid phases and retention in hydration products, the aggressiveness of a pore fluid associated with a blended cement cannot be reliably predicted from the chemical composition of its constituents. It must be evaluated from a laboratory expansion test as suggested by Hobbs (1986). The selection of a reference aggregate for the test was considered in Section 9.4. However, the very low alkali content of silica fume, the low permeability it can produce in a carefully proportioned concrete, and its role as very fine 'sacrificial' silica combine to give this mineral addition a value in inhibiting alkali-silica reaction. Field experience of silica fume concrete in France and Iceland has provided evidence of this (Khayat and Atcin, 1992; Gudmusson and Olafson, 1996).

10. Admixtures and special cements

The development of new cements has been a major part of research activity for most of the twentieth century, driven by the need to obtain cement compositions which improve on Portland cement by providing particular properties such as rapid setting and hardening, improved workability, or increased durability in severe environments. In the past 25 years, the need to reduce energy consumption and, where possible, emissions of the greenhouse gas CO_2, have added to the incentives to introduce new cementitious compositions as well as improved production processes. Since long-term satisfactory performance in use is a major requirement of any new cement, testing to a point where it is widely accepted can involve a prolonged examination of its durability in the environments in which it is to be used.

In this chapter, some special cements are described briefly and, since the properties of Portland cement mortars or concretes can be given special properties by the use of admixtures, a short account of their nature and applications is included.

10.1 Admixtures

The properties of a concrete or mortar containing Portland cement can often be beneficially modified for a particular use by the addition of small amounts of certain chemicals. When the addition is made as the concrete mix is being prepared, the material is described as an admixture. A classification of the commonest admixtures groups them as *accelerators* or *retarders* of set and hardening and *water reducers*, although a particular substance may combine one of the first two characteristics while also reducing the water needed to produce a mix with a given workability. The addition is usually made with the substance in solution to maximise the uniformity of its dispersion, although its introduction may be slightly after the addition of the main bulk of the water where experience has shown that this increases effectiveness. Sensitivity to dose of the admixture is first examined with a sample of the particular cement to be used since the addition required may be influenced by cement fineness and chemistry, in particular the contents of C_3A, free lime, and soluble alkali sulfates. Dose is usually expressed as mass percentage on cement of the active ingredient of an admixture.

10.1.1   Accelerators

These may be employed in precast concrete production or cold weather concreting. They act by increasing the rate of hydration in the acceleratory period, leaving the dormant period largely unaffected. Calcium chloride at 2% is particularly effective in increasing early strength development but the corrosive effect of the chloride ion means that it can not be used in reinforced or prestressed concrete. Calcium formate, nitrate and nitrite are less effective alternatives but no single accelerator is widely accepted (Neville, 1995). It is often more practicable to employ a water-reducing admixture to enhance early strength development.

10.1.2   Retarders

These are valuable in extending the working time of a concrete or mortar in warm conditions since their effect is primarily confined to the dormant period. Hydroxycarboxylic acids (citric acid and those, such as gluconic acid, derived from sugars) and sugars themselves are examples, the latter having drastic effects if an overdose is used. Additions of about 0.25% are usual, but sucrose can extend setting time by as much as 10 h at levels as low as 0.05%. The low dosage of retarders needed is considered to indicate that they function by adsorption on the surfaces of cement grains or, more probably, the hydrates formed initially on them. The poisoning of portlandite (CH) nuclei has also been suggested as a mechanism of retarding set. With cements having a high C_3A content, a greater retardation with a given dose may be obtained by a delay of just two minutes in adding the retarder after the bulk of the water. The initial interaction of C_3A with gypsum is believed to reduce its interaction with the admixture.

Fluorides, phosphates, zinc and lead salts are all retarders which are precipitated from solution (the first two as calcium salts, the second two as hydroxides) as coatings on the surface of cement grains. Phosphate-based admixtures show superior retention of their effectiveness at elevated temperatures (Neville, 1995). Proprietary blends of retarders and plasticisers are employed in ready-mixed mortars with a life of 36–48 h.

10.1.3 Water-reducing (plasticising) admixtures

These enable a reduction of up to 15% in water content to be made while retaining a chosen workability. Usually, a sodium or calcium ligno-sulfonate (by-products of wood pulp manufacture) or a hydroxycarboxylic acid is employed. Sugars in unrefined samples of the former are said to give it an additional function as a retarder. These materials inhibit segregation and can be used in pumped concrete.

Sodium salts of sulfonated naphthalene-formaldehyde co-polymers (PNS) or sulfonated melamine-formaldehyde co-polymers (PMS) are described as *superplasticisers* because they can be used at levels of addition (ca. 0.5%) which make possible water reductions of around 30% without introducing either air entrainment or retardation. The production of very high strength concrete becomes practicable and Neville cites 28-day compressive strengths of $150 \, N/mm^2$ for concrete made at a water/cement ratio of 0.2. Alternatively, the water content may be maintained with no loss of cohesion even with slumps as high as 200 mm. These materials are therefore important in pumpable concrete and self-levelling screeds for flooring. A mix will progressively stiffen (exhibit slump loss) but an additional dose of superplasticiser may be used to prolong working time.

The enhancement of flow by superplasticisers is attributed to their adsorption on initial cement hydration products resulting in an increase in the zeta (ζ) potential at the shear plane of the electrical double layer at the interface of particle and aqueous phase. Bonen and Sarkar (1995) found that the adsorption capacity of a cement depended on the molecular weight of the PNS, the fineness of the cement and its C_3A content, while slump loss was strongly dependent on the ionic strength of the aqueous phase. Cement grains in a paste usually possess a low ζ potential, presumably because of the high ionic strength, but the strongly acidic (dissociated) sulfonate groups (SO_3H) maintain a negative surface charge and ζ potentials in the region of $-30 \, mV$ have been quoted. This would significantly reduce the extent to which particles form agglomerates trapping water which cannot contribute to flow, a factor which is particularly important in concentrated suspensions.

Uchikawa *et al.* (1997) found lower ζ potentials (ca. $-10 \, mV$) in a cement paste to which a PNS co-polymer had been added, although it exhibited satisfactory flow characteristics. They used the relatively recently introduced electrokinetic sonic method, which does not require dilution of a cement paste, and the lower potential could be ascribed to the higher ionic strength of the undiluted suspension. However, the most interesting observation was that a co-polymer of polyacrylic acid and polyacrylic ester was especially effective in increasing paste fluidity, although it did not produce a significant change in the ζ potential of the paste (ca. $-1\,\text{mV}$). They concluded that this was evidence of steric stabilisation. Everett (1988) has provided a useful introduction to the electrical double layer, as well as the possible roles that polymers can play in either coagulating or stabilising colloidal dispersions, depending on their molecular size and structure.

10.1.4   Air entrainment

The value of superplasticisers is attributed to the fact that they can be used at higher doses than the earlier water reducers because they do not induce air entrainment. However, where concrete with superior frost resistance is required, an admixture such as a salt of a fatty acid or an alkali salt of a wood resin is used to enhance air entrainment. To trap air as a dispersion of fine bubbles (foam) in the paste fraction of a concrete, the admixture must first lower the surface tension of the aqueous phase and then stabilise the air bubbles incorporated by the mixer. The latter is achieved by using substances such as long chain fatty acid salts which contain a polar hydrophilic head, for example a sulfonate group, attached to a hydrophobic hydrocarbon chain. The latter resides within the air bubble, while the charged head sits in the aqueous phase and the similar charge distribution around each bubble results in a mutual repulsion which inhibits bubble coalescence. Taylor (1997) pointed out that, in contrast, a superplasticiser such as PNS contains polar groups along the length of the polymer molecule, so that a similar separation of polar and non-polar groups at the air/water interface is not possible. The air bubbles in a hardened concrete provide space for water under pressure from an advancing ice front. Neville (1995) discussed freeze-thaw effects in concrete, an important mechanism of its deterioration in cold, wet conditions.

10.2   Oilwell cements

An oilwell cement is used to seal porous rock formations and to hold the well casing in position. A grout is pumped down the steel casing and forced up the annulus between it and the borehole wall. The two are perforated in the oil production zone. A range of products exists to cater for the range of temperatures and pressures experienced when pumping to different depths. An important property of a grout is, therefore, the rate at which it thickens at elevated temperatures and pressures, since the costs resulting from the loss of a borehole as a result of premature stiffening are considerable. Stiffening is usually examined in a simple laboratory viscometer, the working part containing the grout being held in an autoclave. However, Coveney *et al.* (1996) described a more rapid method of screening cements for use in an oilwell. This involves comparing an infrared reflectance spectrum with those in a database for samples, the thickening times of which have been recorded. Phases such as syngenite ($K_2C\bar{S}_2H$) (the presence of which indicates aeration/ageing of a cement and is linked to stiffening problems) are detected by their characteristic spectra and the acceptability of a sample determined by reference to the database held in a computer.

The American Petroleum Institute classification of cement types, specifications for materials, and test procedures are almost universally employed. Most oilwell cements are based on relatively coarsely ground sulfate-resisting Portland cement clinker. A range of admixtures is employed to give the grout the properties required for specific locations in a well. In addition to gypsum as a set regulator, retarders such as sodium lignosulfonate or gluconic acid are needed at depths of more than about 2000 m. For cementing where temperatures in excess of about 110°C occur, silica is added to prevent the formation of coarse crystals of αC_2SH, which results in an increase in permeability and strength reduction ('retrogression'). Examples of additional admixtures employed are polyacrylamide to reduce water loss in permeable rock zones and weighting agents such as hematite, needed in deep wells to increase grout density. An account of these cements has been given by Bensted (1998).

10.3   Calcium aluminate cement (CAC)

Formerly referred to as high alumina cement (HAC), calcium aluminate cement was developed in France in the early years of the twentieth century to meet several needs: rapid hardening, concreting in cold conditions, sulfate- (sea water) resistance and refractoriness. The major product, *Ciment Fondu*, is manufactured by total melting of a mix of a ferruginous bauxite and limestone at about 1700° in a reverbatory, open hearth furnace. A typical major oxide analysis of the product is: SiO_2 4.5%; Al_2O_3 38%; CaO 38%; Fe_2O_3 10%; FeO 6%. The principal phases present are CA (40–50%) and a ferrite phase (20–40%), the A/F ratio of which varies from grain to grain but is generally less than 1 (Scrivener and Capmas, 1998). Solid solution effects are extensive and minor phases include C_2S and C_2AS. Some $C_{12}A_7$ and a glass may also be present. The minor iron-containing phases present, which include FeO (wüstite), depend on the ratio of Fe^{2+} to Fe^{3+} and they give the clinker its black colour. The cooled, solidified product is ground without addition to a fineness of 300–400 m^2/kg.

Concrete made with Ciment Fondu, unless very vigorously mixed, usually sets more slowly than that made with Portland cement but it can develop the same strength in 1 day as that found in Portland cement concrete in 28 days and the heat generated aids hydration in cold conditions. Strength is mainly derived from the rapid hydration of CA

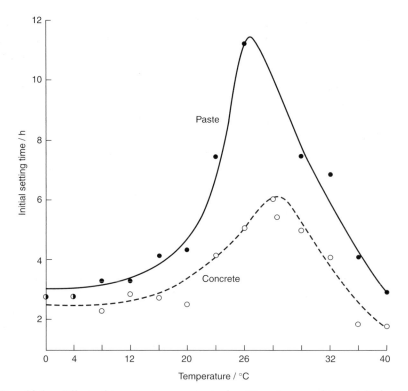

Fig. 10.1. Effect of temperature on typical setting times and initial hydration products for pastes and concretes made with CAC: (A) 5–25° $C_2AH_8 + AH_3$ progressively replace CAH_{10}; (B) 25–35° $C_3AH_6 + AH_3$ progressively replace C_2AH_8. The temperature at which setting time is a maximum is influenced by the $C_{12}A_7/CA$ ratio in the cement (Scrivener and Capmas, 1998)

which occurs by a process of dissolution followed by nucleation and precipitation of a crystalline hydrate. The hydrate precipitated depends on the temperature of hydration as shown in Fig. 10.1, in which the effect of temperature on setting time is shown. If $C_{12}A_7$ is present, C_2AH_8 is formed even at the lower temperatures. The hydration of calcium aluminate cements involves the conversion of Al from the 4- coordination in the anhydrous phases to the 6- coordination present in the crystalline hydrates and AH_3, so that it can be followed by ^{27}Al NMR (Kirkpatrick and Cong, 1993).

The rapidly formed initial product in the hydration of CA at low temperatures is CAH_{10} which is metastable; it undergoes 'conversion' by a solid state process which is accelerated by warm, humid conditions:

$$3CAH_{10} \longrightarrow C_3AH_6 + 2AH_3 + 18H \qquad (10.1)$$

where AH_3 is an aluminium hydroxide gel which subsequently crystallises to form gibbsite. However, following the conversion reaction by synchrotron radiation energy dispersive diffraction, Rashid *et al.* (1992) found that in the temperature range 70–90°, C_2AH_8 was an initial, transitory intermediate. In a later investigation (1994) of the reaction at 50°, they found that both crystal forms of this phase were formed sequentially. The direct conversion to C_3AH_6 was only observed after its nucleation by the indirect route.

The reaction in Equation (10.1) is accompanied by a significant increase in the density of the solid phases and, consequently, a marked increase in the porosity of the concrete and reduction in its strength result, although some of the water formed may react with remaining anhydrous cement to give a modest subsequent increase. The design of a structure can be based on the converted strength and the value for this can be optimised by using a water/cement ratio of less than 0.4. This produces a concrete in which the reduction in strength is not too severe but ensuring adequate strength by use of such low water contents has been considered too difficult for normal working conditions (Neville, 1995). In a humid environment serious problems have also arisen where CAC concrete structural members have been in contact with Portland cement concrete. Loss of strength in the former is the result of alumina leaching from the CAC by soluble alkali from the Portland cement (*alkaline hydrolysis*) and loss of integrity in the concrete is accelerated by simultaneous carbonation.

Calcium aluminate cements are now primarily used in the UK for non-load bearing and refractory applications. The more refractory cream and white grades are produced by a clinkering process using low iron content bauxites, or gibbsite from the Bayer process, and either a pure limestone or quick lime. They may contain CA_2, CA_6, and alumina itself, as the proportion of the latter is increased in the raw mix to raise the refractoriness of the product.

10.4   Alkali-activated slag and aluminosilicate cements

The latent hydraulicity of blastfurnace slag can be activated by alkali as well as lime and a wide range of siliceous slags, from the metallurgical and phosphorous producing industries, can be used in *alkali-activated slag cements*. Apart from the phosphorus slag which contains only a low level of alumina, the principal components of the slags used are typically: CaO (35–45%); SiO_2 (30–40%); Al_2O_3 (10–25%); MgO (1–8%) and a small amount of iron oxide (FeO). A number of sources of the alkali have been used, such as their hydroxides and carbonates, but sodium silicate (water glass) has been found to give good strength development and is used at a level of 3–6% Na_2O in the mix.

These materials were patented by Glukhovsky in 1958 and have since been increasingly used in precast concrete products, in Eastern Europe, Russia and China, for example. Low cost is a major advantage resulting from the incorporation of a waste in a low energy product. Other advantages include high early strength potential, low heat of hydration and good resistance to sulfate and chloride ions. Disadvantages include rapid slump loss and setting, a more complicated gauging procedure than is needed with Portland cement, and potential alkali-aggregate reaction.

Richardson *et al.* (1994) examined the hydration products of ground granulated blastfurnace slag activated by 5M KOH solution, using TEM and NMR. The major hydration products were a semicrystalline C–S–H with a C/S ratio around one for both inner and outer product. A magnesium-rich phase (Mg/Al ca. 2.5) apparently related to hydrotalcite was intimately mixed with the inner product. Energy loss spectroscopy in the TEM supported the idea that aluminium was present in the C–S–H, substituting for bridging silicon in the dreierkette type silicate chain (Fig. 7.10).

Supersulfated cement can be classified as a slag cement activated in an alkaline environment because a small amount of Portland cement (ca. 5%) is used to activate 80–85% blastfurnace slag and 10–15% of anhydrite; C–S–H and ettringite are the main hydration products. Supersulfated cements are covered by BS 4248: 1974 but are not currently manufactured in the UK.

A cementitious material can also be made by *alkali activation of aluminosilicates*. The latter may either be a natural clay (cement or lime in soil stabilisation, for example) or a kaolinitic clay dehydrated at 750° to produce metakaolin. The latter is either reacted with an alkali polysilicate solution or mixed with water after dry blending with the polysilicate. Davidovits (1987) referred to these materials as *geopolymer cements*. Hydration is rapid at room temperature with strengths of $20 \, \text{N/mm}^2$ reached after 4 h and $70–100 \, \text{N/mm}^2$ after 28 days (Davidovits, 1994). Malek and Roy (1997) examined a range of alkali/alkaline earth-aluminosilicate compositions and considered that freedom from alkali-silica reaction in concretes containing them was explained by the fixing of the alkali metal cations in an aluminosilicate framework structure.

10.5 Calcium sulfoaluminate cements

The anhydrous calcium sulfoaluminate $C_4A_3\bar{S}$ (frequently referred to as Klein's compound although first described in 1957 by Ragozina) is the key ingredient of an increasing number of cements with valuable properties. Its importance derives from the fact that on hydration in the presence of lime and calcium sulfate, it rapidly hydrates to form ettringite.

Advantages follow not only from the rapidity of the formation of this hydration product but also from its acicular or prismatic crystal morphology. It has been employed in a range of compositions under conditions taking advantage of one or more of the features of its formation: high early strength development (levels of over $50 \, \text{N/mm}^2$ in 12–24 h are possible); expansion; and an ability to solidify high water content slurries.

In this section, reference is also made to cements which can be described as low energy materials since significant savings follow, not only from the lower temperature and total heat input needed to produce $C_4A_3\bar{S}$, but also from the fact that the considerable densification resulting from clinkering is avoided; grinding a relatively friable product of solid state reactions requires far less energy. Note, however, that the unground high temperature product is usually referred to as a clinker. Further economies are possible since some of these cements can be made from limestone and waste materials, where these are available with the requisite amounts of aluminium and iron. An introduction to these cements, including mining applications, has been given by Sharp *et al.* (1999).

10.5.1 *Expansive and shrinkage compensated cements*

For one commercial product in the USA, $C_4A_3\bar{S}$ clinker is produced from a mix of limestone, an aluminous material such as bauxite, and gypsum or anhydrite, at temperatures at or above 1300°. The expansive hydration reaction involves calcium hydroxide, provided by the simultaneous hydration of a Portland clinker in a composite cement:

$$C_4A_3\bar{S} + 8C\bar{S}H_2 + 6CH + 74H \longrightarrow 3C_6A\bar{S}_3H_{32} \qquad (10.2)$$

The important determinants of the performance of a cement in the restrained expansion of a concrete are the ratios of lime (Portland cement) and calcium sulphate to $C_4A_3\bar{S}$ and an assessment of the potential of a cement formulation can be obtained by measuring its unrestrained expansion. Following the crystal growth pressure hypothesis, the high pH produced by lime results in a high supersaturation and the rapid, expansive crystallisation of ettringite close to the surface of the hydrating sulfoaluminate. In the absence of lime, a lower supersaturation leads to the formation of larger crystals of ettringite at regions remote from the anhydrous sulfoaluminate surface (Section 6.6).

The sulfoaluminate clinker is either interground with Portland clinker or they are separately ground and then blended, the latter having the advantage that a coarser sulfoaluminate can be employed when a delayed expansive effect is desired. Commercial compositions are either designed to offset the ultimate drying shrinkage of a concrete or to develop a net expansive force, which under the restraint of tensioned wires embedded in a concrete, results in self-stressing (Neville, 1995). A range of compositions based on $C_4A_3\bar{S}$ is produced in the USA and designated type K.

Brown (1998) described a number of cements, manufactured in the UK, based on a clinker containing Klein's compound. Produced in a slightly modified conventional rotary kiln from relatively pure aluminous materials, it has an estimated compound composition: $C_4A_3\bar{S}$- 57%; CA- 17%; C_2AS- 17%. Desired properties for a particular application are obtained by blending the ground clinker with Portland cement and varying the anhydrite to $C_4A_3\bar{S}$ ratio. At higher ratios an expansive (shrinkage compensated) cement similar to type K is obtained. Hydration results in a lower portlandite content than in hardened Portland cement paste but the pH is high enough to stabilise ettringite (>10.5) while having the advantage of not being aggressive to glass fibre reinforcement. Scanning electron microscopy showed that the network of acicular ettringite crystals, providing early strength in hydration, is infilled by C–S–H produced from the Portland cement. Most of the applications have involved low permeability cement-rich mixes and satisfactory durability has been found in use to date (12 years). In a mining application, a slurry of the ground clinker is splash-mixed with one of anhydrite and calcium hydroxide. The mix, with a water/cement ratio of about 2, rapidly provides sufficient strength (ca. 2–3 N/mm^2) to be used in a temporary support for the tunnel walls behind a coal-cutting machine.

10.5.2 Sulfoaluminate-belite cements

In a Portland cement concrete, belite makes a contribution to long-term strength only (after 14–28 days), assuming that it is held at an adequate humidity (>ca. 80%) for that time. Efforts have, therefore, been made over a long period to develop an active (more rapidly hydrating) form of βC_2S so that a raw mix would not need the high LSF required for current Portland cement manufacture. This would have the advantage of significantly reducing the energy used in burning the clinker to an acceptable free lime as well as reducing the emission of CO_2. Methods of activation considered have included rapid cooling of a clinker and the incorporation of impurity ions in solid solution in the C_2S. A high surface area, particularly reactive, form of βC_2S can be prepared by the dehydration of hillebrandite ($Ca_2SiO_3(OH)_2$) (Okada et al., 1994). The latter is prepared hydrothermally from lime and silica but the process is not of commercial interest.

Table 10.1.   Composition (mass per cent) of sulfoaluminate clinkers (Wang and Su, 1994)

Type	Fe_2O_3	$C_4A_3\bar{S}$	C_2S	C_4AF
Low iron	1–3	55–75	15–30	3–6
High iron	15–30	35–55	15–35	15–30
Sudoh*	1.6	54	32	5

* Clinker with a free lime of 9% described by Sudoh *et al.* (1980).

A more practical method of increasing the utilisation of belite involves making a clinker in the C–S–A–F–$\bar{S}$ system in which, between 1280° and 1400°, the major phases formed are $C_4A_3\bar{S}$ and C_2S, plus a ferrite phase if ferruginous raw materials such as red bauxites are employed. High early strength results from the hydration of the $C_4A_3\bar{S}$ as it reacts with anhydrite or gypsum (Equation (10.2)) supplemented by the slowly hydrating belite as the concrete matures.

Cements of this type have been manufactured in China since the 1970s, with annual production passing the 1 000 000 t level in the mid-1990s. A range of applications is made possible by varying the calcium sulfate to $C_4A_3\bar{S}$ molar ratio from 0.5 to 6, since this determines the potential expansion accompanying hydration. At the lowest ratio, a rapid hardening cement for normal concrete is produced while at the highest, potential expansion is sufficient for the cement to be used in prestressed concrete. The range of compositions manufactured was extended during the 1980s to cements with higher ferrite contents (Table 10.1). Zhang *et al.* (1999) have given an account of the properties and uses of these cements in China, including some aspects of their durability. They reported good resistance to frost, sulfate and sea water. The low permeability these cements can produce in concrete is considered to provide adequate protection of steel reinforcement at the lower pH values obtaining (<12), although some initial oxidation of it has been observed.

Concerns have been expressed about the durability of calcium sulfoaluminate cements, since in moist conditions ettringite is carbonated to form calcium carbonate and aluminium hydroxide gel which might be expected to lead to a strength reduction. This was observed in accelerated carbonation tests, although the effect was reversed by the incorporation of C_4AF in the experimental cements examined (Beretka *et al.*, 1997). Similar observations have been made by others but experience of over 30 years with field concretes has not raised serious problems. In an examination of plain and sulfoaluminate concrete samples taken from buildings in Japan, Makai *et al.* (1993) found that carbonation depths were similar in both and that C–S–H and CH had been carbonated to a greater extent than ettringite.

10.5.3   Practical considerations

Calcium sulfoaluminate cements have been described as 'cements for the future'. However, a problem arises with any new product in balancing production and demand growing from a low start. In a modern plant, production of Portland cement is very high, 1 million tonnes per annum being not unusual. This is the case in many countries where modernisation has replaced most if not all small kilns. Furthermore, in many emerging economies large plants have been installed where a construction industry was developing to a level providing an adequate regional market. The problem is compounded if a product requires a long period of durability assessment and test marketing to gain general acceptance by the construction and concrete products industries. The production of two very different clinkers in one large kiln would be inefficient, since it would require purging of the whole system at times of changeover and much increased wear of refractories. Despite these considerations, calcium sulfoaluminate cements have now been used for more than 30 years and have several national standards relating to their composition and use.

Appendix. Pore size distribution from an adsorption isotherm

Liquid adsorbate is stable in narrow pores at pressures below its saturated vapour pressure (p_o) and this phenomenon is known as capillary condensation. The critical pressure (p) for liquid stability in a cylindrical pore is related to the radius of the meniscus (r) which is equal to the pore radius, by the Kelvin equation (A1); if the pressure falls below this value the liquid evaporates.

$$\ln p/p_0 = \frac{-2\gamma V}{rRT} \cos \phi \qquad (A1)$$

in which, γ, V and ϕ are respectively the surface tension, molar volume and wetting angle of the liquid; R and T have the usual meaning.

The Kelvin equation is usually applied to data from the desorption branch of an isotherm and r is replaced by $r - t$, where t is the thickness of the adsorbed film remaining on the walls of the pore after the liquid has evaporated. The so-called 't' curve (t against p/p_o) is obtained from adsorption measurements with the same adsorbate on a non-porous material having a surface as similar chemically as possible to the sample under examination (Gregg and Sing, 1982).

The principal problem in using this method for hcp is that while Equation (A1) can be modified for the more probable slit-shaped pore, the pore system in hcp must be more complex than a set of slits differing only in width. Variations in width along the length of each pore giving rise to wedge-shaped portions and to constrictions would be expected. Furthermore, the maximum at around 2–3 nm in the curves in Fig. 8.7 (corresponding to a steep desorption region such as that at DA in Fig. 8.6) is common and found with many other porous adsorbents. Such a step in the desorption branch of an isotherm has been ascribed to the existence of either narrow entrances to larger pores (C in Fig. 8.14) or constrictions within a system of pores of variable width, and it leads to an exaggeration of the volume of pores with the corresponding Kelvin radius. Another explanation given for abrupt desorption is that the decrease in pressure has, at this point, increased the tensile stress in the capillary condensed liquid to a value greater than its tensile strength so that it suddenly evaporates (Everett, 1967).

Suggested further reading

Abbreviations   ACI SP American Concrete Institute Special Publication
                IC(S)CC International Congress (Symposium) on the Chemistry of Cement
                RILEM   International Union of Testing and Research Laboratories for Materials and Structures
                VDZ   Association of German Cement Industries
                ZKG International   formerly Zement Kalk Gips

Chapter 1
1.1   Hewlett, P.C. (ed.) (1998). *Lea's chemistry of cement and concrete*. 4th edn, 1053 pp, Arnold, London.
1.2   Macphee, D.E. and Lachowski, E.E. (1998). Cement components and their phase relations. Ibid, 95–129.
1.3   Taylor, H.F.W. (1997). *Cement Chemistry*. 2nd edn, 459 pp, Thomas Telford Publishing, London.

Chapter 2
2.1   Fundal, E. (1996). Burnability of cement raw meal with matrix correction, *World Cement*, **27**, No. 4, 63–68.
2.2   Jackson, P.J. (1998). Portland cement: classification and manufacture. *Lea's chemistry of cement and concrete*. 4th edn, Arnold, London, 25–94.
2.3   Manning, D.A.C. (1995). *Industrial minerals*, Chapman and Hall, London.
2.4   Moir, G.K. (1997). Influence of raw mix heterogeneity on ease of combination and clinker strength potential. *10th ICCC, Gothenburg*. Paper 1i041.
2.5   van Olphen, H. (1977). *An introduction to clay colloid chemistry*. 2nd edn, Interscience, London.
2.6   Tucker, E. (1981). *Sedimentary petrology: an introduction*. Blackwell Scientific, Oxford.

Chapter 3
3.1   Barry, T.I. and Glasser, F.P. (1999). Calculations of Portland cement clinkering reactions. *Adv. Cem. Res.*, at press.
3.2   Erhard, H.S. and Scheuer, A. (1994). Burning technology and thermal economy. *1993 VDZ Congress: process technology of cement manufacturing*. Bauverlag, Wiesbaden, 1994, 278–295.
3.3   Glasser, F.P. (1998). The burning of Portland cement. *Lea's chemistry of cement and concrete*. 4th edn, Arnold, London, 185–240.
3.4   Johansen, V. (1989). Cement production and cement quality. *Materials science of concrete I*. American Ceramic Society, Westerville, Ohio, 27–72.

3.5   Jøns, E. and Hundebøl, S. (1995). Prediction of heat of reaction of cement raw meal. *ZKG International*, **48**, No. 9, 453–462.

3.6   Kerton, C.P. (1994). (Clift, R. and Seville, J.P.K. eds). Behaviour of volatile materials in cement kiln systems. *Gas cleaning at high temperatures*. Blackie Academic, London, 589–603.

3.7   Kraisha, Y.H. and Dugwell, D.R. (1992). Coal combustion and limestone calcination in a suspension reactor. *Chem. Eng. Sci.*, **47**, No. 5, 993–1006.

3.8   Lowes, T.M. and Evans, L.P. (1994). The effect of burner design and operating parameters on flame shape, heat transfer and NOx and SO_3 cycles, 305–314.

3.9   Moir, G.K. (1997). Cement production: state of the art. *Brit. Ceram. Trans.*, **96**, No. 5, 204–212.

3.10   Seidel, G., Huckhauf, H. and Stark, J. (1980). *Technology of building materials — burning processes and plant*. VEB, Berlin and Editions Septima, Paris.

3.11   Timashev, V.V. (1980). The kinetics of clinker formation. *7th ICCC, Paris*, 1 I-3 1–20. Editions Septima, Paris.

Chapter 4

4.1a   Aldridge, L.P. (1982). Accuracy and precision of phase analysis in Portland cement by Bogue, microscopic and X-ray diffraction methods. *Cem. Concr. Res.*, **12**, 381–398.

4.1b   Campbell, D.H. (1986). *Microscopical examination and interpretation of Portland cement and clinker*. Portland Cement Association, Skokie, Illinois.

4.2   Donald, A. (1998). Taking SEMs into a new environment. *Materials World — J. Inst. Mat.*, **6**, July, 399–401.

4.3a   Groves, G.W. (1981). Portland cement clinker viewed by transmission electron microscopy. *J. Mat. Sci.*, **16**, 1063–1070.

4.3b   Groves, G.W. (1982). Twinning in beta-dicalcium silicate. *Cem. and Concr. Res.*, **12**, No. 5, 619–624.

4.4a   Maki, I. (1994). (Gartner, E.M. and Uchikawa, H. eds) Processing conditions of Portland cement clinker as viewed from the fine textures of the constituent minerals. *Cement technology. Ceramic Trans.*, **40**, 3–17. American Ceramic Society, Westerville, Ohio.

4.4b   Maki, I. *et al.* (1995). Anisotropic light absorption of calcium aluminoferrite phase in Portland cement clinker. *Cem. and Concr. Res.*, **25**, No. 4, 863–869.

4.5   Möller, H. (1998). Automatic profile investigation by the Rietveld method for standardless quantitative phase analysis. *ZKG International*, **51**, No. 1, 40–50.

4.6   Scrivener, K.L. (1997). Microscopy methods in cement and concrete science. *World Cement*, **28**, No. 9, 92–112.

4.7   Scrivener, K.L. and Taylor, H.F.W. (1995). Microstructural and microanalytical study of strongly reduced Portland cement clinker. *World Cement*, **26**, No. 8, 34–44.

4.8   Taylor, J.C. and Aldridge, L.P. (1993). Phase analysis of Portland cement by full profile standardless quantitative X-ray diffraction: accuracy and precision. *Adv. in X-ray Analysis*, **36**, 309–314.

4.9   Young, R.A. (1993) (Young, R.A. ed). Introduction to the Rietveld method. *The Rietveld method*. International Union Crystallography, OUP, Oxford.

Chapter 5

5.1   Allen, T. (1996). *Particle size measurement: Volume 1 — Powder sampling and particle size measurement. Volume 2 — Surface area and pore size determination*. Chapman and Hall, London.

5.2   Bonen, D. and Diamond, S. (1991). Application of image analysis to a comparison of ball mill and high pressure roller mill ground cement. *Proc. 13th Int. Conf. on Cement Microscopy*, International Cement Microscopy Association, Duncanville, Texas, 101–119.

5.3   Ellerbrock, H.G. and Mathiak, H. (1994). Comminution technology and energy management. *1993 VDZ Congress: progress technology of cement manufacturing*. Bauverlag, Wiesbaden, 1994, 630–647.

5.4   Geurts, R., van Loo, W. and Sumner, M. (1992). The performance of laser granulometers in the cement industry. *World Cement*, **23**, No. 5, 51–53.

5.5   Kuhlmann, K., Ellerbrock, H.G. and Sprung, S. (1985). Particle size distribution and properties of cement: Part 1 — Strength of Portland cement. *Zement Kalk Gips*, **4/85**, 169–178 (translation **6/85**, 136–144).

5.6   Maki, I. *et al.* (1993). Grindability and textures of alite and belite. *Cem. Concr. Res.*, **23**, 1078–1084.

5.7   Odler, I. and Chen, Y. (1995). Influence of method of comminution on the properties of cement. *ZKG International*, **9/95**, 496–500.

5.8   Osbaeck, B. and Johansen, V. (1989). Particle size distribution and rate of strength development of Portland cement. *J. Am. Ceram. Soc.*, **72**, No. 2, 197–201.

5.9   Scrivener, K.L. (1987). The microstructure of anhydrous cement and its effect on hydration. *Mat. Res. Soc. Proc.*, **85** 39–46.

Chapter 6

6.1   Banfill, P.F.G. (1996). (Bartos, P.J.M. *et al.* eds). Vibration and rheology of fresh concrete. *Production methods and workability of concrete RILEM Proc.* **32**, E & FN Spon, London, 319–326.

6.2   British Standards Institution. Methods of testing cement BS EN 196. (Parts published from 1989 as prepared.)

6.3   Brookbanks, P. (1994) (Dhir, R.K. and Jones, M.R. eds). Conformity criteria for common European cements. *Eurocements: impact of ENV 197 on concrete construction*. E & FN Spon, London, 183–193.

6.4   Hobbs, D.W. ed. (1998). *Minimum requirements for durable concrete*. British Cement Association, Crowthorne, Berks.

6.5   de Larrard *et al.* (1994). (Bartos, P.J.M. ed.). Design of a rheometer for fluid concretes. *Special concretes: workability and mixing. RILEM Proc.* **24**, E & FN Spon, London, 201–208.

6.6   Lawrence, C.D. (1998). Physicochemical and mechanical properties of Portland cements. *Lea's chemistry of cement and concrete*. 4th edn, Arnold, London, 343–419.

6.7   Neville, A.M. (1995). *Properties of concrete*. 4th edn. Longmans, Harlow, Essex.

6.8   Tattersall, G.H. (1991). *Workability and quality control of concrete*. E & FN Spon, London.

6.9   Taylor, M.G. (1994). Key test methods and limits. *Eurocements: impact of ENV 197 on concrete construction*. E & FN Spon, London, 69–98.

Chapter 7

7.1   Barnes, P. (1996). Applied crystallography solutions to problems in solid state chemistry — case examples with ceramics, cements and zeolites. *J. Chem. Soc. Faraday Trans.*, **92**, No. 12, 2187–2196.

7.2   Bensted, J. (1987). Some applications of conduction calorimetry to cement hydration. *Adv. Cem. Res.*, **1**, No. 1, 35–44.

7.3   Brough *et al.* (1994). In situ solid state NMR studies of Ca_3SiO_5: hydration at room temperature and at elevated temperatures using ^{29}Si enrichment. *J. Mat. Sci.*, **29**, 3926–3940.

7.4   Diamond, S. (1996). Delayed ettringite formation — processes and problems. *Cem. & Concr. Composites*, **18**, 205–215.

7.5   Glasser, F.P. (1992). (Nonat, A. and Mutin, J.C. eds). Co-existing solids and the aqueous phase in Portland cement. *Hydration and setting of cements, Proc. Int. RILEM workshop 1991*, E & FN Spon, London, 101–123.

7.6   Granger, P. (1994). (Colombet, P. and Grimmer, A-R. eds). Nuclear magnetic resonance basic principles. *Application of NMR spectroscopy to cement science.* Gordon & Breach, Reading, Berks, 5–28.

7.7   Jennings, H.M. *et al.* (1997). Phase diagrams relevant to the hydration of C_3S. Part 1 — A case for metastable equilibrium. Part 2 — A phase diagram for the CaO–SiO_2–H_2O system. *10th ICCC Gothenburg, 1997.* Papers 2ii057 and 2ii058.

7.8   Jupe, A.C. *et al.* (1996). Fast in situ X-ray diffraction studies of chemical reactions: a synchrotron view of the hydration of tricalcium aluminate. *Phys. Rev. B. Condensed Matter*, **53**, No. 22, R14, 697–700.

7.9   Kirkpatrick, R.J. and Cong, X-D. An introduction to ^{27}Al and ^{29}Si NMR spectroscopy of cements and concretes. *Application of NMR spectroscopy to cement science.* Gordon & Breach, Reading, Berks, 55–75.

7.10   Lawrence, C.D. (1995). Delayed ettringite formation — an issue? *Mat. Sci. Concr.*, **IV**, 113–154.

7.11   Parrott, L.J. *et al.* (1990). Monitoring Portland cement hydration: comparison of methods. *Cem. Concr. Res.*, **20**, No. 6, 919–926.

7.12   Skibsted, J. *et al.* (1997). Hydration kinetics for the alite, belite and calcium aluminate phases in Portland cements from ^{27}Al and ^{29}Si MAS NMR spectroscopy. *10th ICCC, Gothenburg, 1997.* Paper 2ii056.

7.13   Wieker, W. *et al.* (1997). Recent results of solid state NMR investigations and their possibilities of use in cement chemistry. *10th ICCC Gothenburg, 1997.* **1** Plenary lecture 2.

7.14   Zelwer, A. (1973). Electrochemical study of the aqueous phase in the hydration of C_3S. *Rev. des Materiaux de Construction et de Travaux Publics*, (681), 20.

Chapter 8

8.1   Beaudoin, J.J. and Brown, P.W. (1992). The structure of hardened cement paste. *9th ICCC, New Dehli, 1992*, **1**, 485–525.

8.2   Bentz, D.P., Schlangen, E. and Garboczi, E.J. (1995). Computer simulation of interfacial zone microstructure and its effect on the properties of cement-based composites. *Mat. Sci. Concr.*, **IV**, 155–199.

8.3   Bentz, D.P. (1997). Three-dimensional computer simulation of Portland cement hydration and microstructure development. *J. Am. Ceram. Soc*, **80**, No. 1, 3–21.

8.4   Bonen, D. and Diamond, S. (1994). Interpretation of compositional patterns found by quantitative energy dispersive X-ray analysis for cement paste constituents. *J. Am. Ceram. Soc.*, **77**, No. 7, 1875–1882.

8.5   Brown, P.W., Shi, D. and Skalny, J. (1991). Porosity/permeability relationships. *Mat. Sci. Concr.*, **II**, 83–109.

8.6   Darwin, D. and Abou-Zeid, M.N. (1995). Application of automated image analysis to the study of cement paste microstructure. *Mat. Res. Soc. Symp. Proc.*, **370**, 3–12.

8.7   Diamond, S. and Leeman, M.E. (1995). Pore size distributions in hardened cement paste by SEM image analysis. *Mat. Res. Soc. Symp. Proc.*, **370**, 217–226.

8.8   Fagerlund, G. (1974). Strength and porosity of concrete. *Proc. RILEM/IUPAC Symp.: pore structure and properties of materials*, D-51-73, Academia Prague.

8.9   Garboczi, E.J. and Bentz, D.P. (1991). Fundamental computer simulation models for cement-based materials. *Mat. Sci. Concr.*, **II**, 249–277.

8.10   Harrisson, A.M., Winter, N.B. and Taylor, H.F.W. (1986). An examination of some pure and composite Portland cement pastes using SEM with X-ray analytical facility. *8th ICCC, Rio de Janeiro, 1986*, **4**, 170–175.

8.11   Hooton, R.D. (1989). What is needed in a permeability test for the evaluation of concrete quality. *Mat. Res. Soc. Symp. Proc.*, **137**, 141–149.

8.12   Jennings, H.M. and Tennis, P.D. (1994). Model for the developing microstructure in Portland cement pastes. *J. Am. Ceram. Soc.*, **77**, No. 12, 3161–3172.

8.13   Rarick, R.L., Bhatty, J.I. and Jennings, H.M. (1995). Surface area measurement using gas sorption: application to cement paste. *Mat. Sci. Concr.*, **IV**, 1–39.

8.14   Richardson, I.G. and Groves, G.W. (1993). Microstructure and microanalysis of hardened ordinary Portland cement pastes. *J. Mat. Sci.*, **28**, 265–277.

8.15   Scrivener, K.L. and Pratt, P.L. (1996) (Maso, J.C. ed.). Characterisation of interfacial microstructure. RILEM: *Interfacial transition zone in concrete*. E & FN Spon, London, 3–17.

8.16   Viehland, D. *et al.* (1996). Mesostructure of calcium silicate hydrate (C–S–H) gels in Portland cement paste: short range ordering, nanocrystallinity and local compositional order. *J. Am. Ceram. Soc.*, **79**, No. 7, 1731–1744.

8.17   Xi, Y. and Jennings, H.M. (1993). Relationships between microstructure and creep and shrinkage of cement paste. *Mat. Sci. Concr.*, **III**, 37–70.

8.18   Young, J.F. (1993). Editorial: Macro-defect free cement: a novel composite material. *J. Mat. Civ. Eng.*, **5**. No. 4, 423–426 (14 references).

Chapter 9

9.1   American Concrete Institute (1994). Use of natural pozzolans in concrete. *ACI Committee Report 232.1 R-94*. Detroit, Michigan.

9.2   Bijen, J. and Pietersen, H. (Grutzek, M.W. and Sarkar, S.L. eds) (1994). Mineral admixtures: reactions, microstructure and macro-properties. *Advances in cement and concrete*. Am. Soc. Civ. Engrs, 292–328.

9.3   Brough, A.R. *et al.* (1997). ^{29}Si enrichment and selective enrichment for study of the hydration of model cements and blended cements. *10th ICCC, Gothenburg, 1997*. Paper 3v001.

9.4a   Building Research Establishment (1997). *Alkali-silica reaction in concrete*. Digest 330. Watford, UK.

9.4b   Building Research Establishment (1996). *Sulfate and acid resistance of concrete in the ground*. Digest 363. Watford, UK.

9.5   Coole, M.J. (1988). Heat release characteristics of concrete containing ground granulated blastfurnace slag in simulated large pours. *Mag. Concr. Res.*, **40**, No. 144, Sept. 152–158.

9.6   Crammond, N.J. and Halliwell, M.A. (1997). (Scrivener, K.L. and Young, J.F. eds). Assessment of the conditions required for the thaumasite form of sulphate attack. *Mechanisms of chemical degradation of cement-based systems*. E & FN Spon, London, 193–200.

9.7   Dalziel, J.A. and Gutteridge, W.A. (1986). The influence of pulverised fuel ash on the hydration and certain physical properties of Portland cement paste. *C&CA Technical Report 560*. British Cement Association, Crowthorne, Berks.

9.8   Dhir, R.K. (1994). Additional materials and allowable contents in cement. *Eurocements: impact of ENV 197 on concrete construction*. E & FN Spon, London, 57–68.

9.9   Fidjestol, P. and Lewis, R. (1998). Microsilica as an addition. *Lea's chemistry of cement and concrete*. 4th edn, Arnold, London, 675–708.

9.10   Frearson, J.P.H. and Higgins, D.D. (1995) (Malhotra, V.M. ed.). Effect of test procedures on the assessment of sulfate resistance of slag cements. *ACI SP*, **153**, 975–993.

9.11   Helmuth, R. and Stark, D. (1992). Alkali-silica reactivity mechanisms. *Mat. Sci. Concr.*, **III**, 131–208.

9.12   Hobbs, D.W. (1994). The effectiveness of PFA in reducing the risk of cracking due to ASR in concretes containing cristobalite. *Mag. Concr. Res.*, **46**, No. 168, Sept., 167–175.

9.13   Hobbs, D.W. (1988). *Alkali-silica reaction in concrete*. Thomas Telford, London.

9.14   Idorn, G.M. Johansen, V. and Thaulow, N. (1992). Assessment of causes of cracking in concrete. *Mat. Sci. Concr.*, **III**, 71–104.

9.15   Justnes, H., Sellevold, E.J. and Lundevall, G. (1993). High strength concrete binders. Part A: reactivity and composition of cement pastes with and without condensed silica fume. *ACI SP 132*, **2**, 873–889 (Part B, 891–902).

9.16   Lachowski, E.E. *et al.* (1997). Compositional development (solid and aqueous phase) in aged slag and fly ash blended cement pastes. *10th ICCC, Gothenburg, 1997*. Paper 3ii091.

9.17   Lawrence, C.D. (1990). Sulphate attack on concrete. *Mag. Concr. Res.*, **42**. No. 153, Dec., 249–264.

9.18   Lawrence, C.D. (1992). Influence of binder type on sulphate resistance. *Cem. Concr. Res.*, **22**, No. 6, 1047–1058.

9.19   Lumley, J.S. *et al.* (1996). Degrees of reaction of the slag in some blends with Portland cements. *Cem. Concr. Res.*, **26**, No. 1, 139–151.

9.20   Massazza, F. (1998). Pozzolana and pozzolanic cements. *Lea's chemistry of cement and concrete*. 4th edn, Arnold, London, 471–631.

9.21   Mehta, P.K. (1992). Sulfate attack on concrete — a critical review. *Mat. Sci. Concr.*, **III**, 105–130.

9.22   Moranville-Regourd, M. (1998). Cements made from blastfurnace slag. *Lea's chemistry of cement and concrete*. 4th edn, Arnold, London, 633–674.

9.23   Neville, A.M. (1995). *Properties of concrete*. 4th edn. Longmans, Harlow, Essex.

9.24   Patel, H.H., Pratt, P.L. and Parrott, L.J. (1989). Porosity in the microstructure of blended cements containing fly ash. *Mat. Res. Soc. Symp. Proc.*, **136**, 233–242.

9.25   Richardson, I.G. (1997). The structure of C–S–H in hardened slag cement pastes. *10th ICCC, Gothenburg, 1997*. Paper 2ii 068.

9.26   RILEM (1991). (Wesche K. ed.). *Fly ash in concrete: Properties and performance*. E & FN Spon, London.

9.27   Taylor, H.F.W. and Gollop, R.S. (1997). (Scrivener, K.L. and Young, J.F. eds). Some chemical and microstructural aspects of concrete durability. *Mechanisms of chemical degradation of cement-based systems*. E & FN Spon, London, 177–184.

9.28   Tikalsky, P.J. and Carrasquillo, R.L. (1993). Fly ash evaluation and selection for use in sulfate-resistant concrete. *ACI Matt. J.*, 545–551.

Chapter 10

10.1   Bensted, J. (1998). Special cements. *Lea's chemistry of cement and concrete*. 4th edn, Arnold, London, 779–836 (includes oilwell cements).

10.2   Cong, X. and Kirkpatrick, R.J. (1993). Hydration of calcium aluminate cements: a solid state ^{27}Al NMR study. *J. Am. Ceram. Soc.*, **76**, No. 2, 409–416.

10.3   Edmeades, R.M. and Hewlett, P.C. (1998). Cement admixtures. *Lea's chemistry of cement and concrete*. 4th edn, Arnold, London, 837–901.

10.4   Everett, D.H. (1988). *Basic principles of colloid science*. Royal Soc. Chem. (Paperbacks Series) London.

10.5   Fletcher, P. and Coveney, P. (1995). Prediction of thickening times of oil field cements using artificial neural networks and Fourier transform infrared spectroscopy. *Adv. Cem. Based Mat.*, **2**, No. 1, 21–29.

10.6   Lan, W. and Glasser, F.P. (1996). Hydration of calcium sulfoaluminate cements. *Adv. Cem. Res.*, **8**, No. 31, 127–134.

10.7   Lawrence, C.D. (1998). The production of low energy cements. *Lea's chemistry of cement and concrete*. 4th edn, Arnold, London, 421–470.

10.8   Rashid, S. *et al.* (1994). Conversion of calcium aluminate cement hydrates re-examined with synchrotron energy dispersive diffraction. *J. Mat. Sci. Letters*, **13**, 1232–1234.

10.9   Richardson, I.G. *et al.* (1994). The characterisation of hardened alkali-activated blastfurnace slag pastes and the nature of the calcium silicate hydrate (C–S–H) phase. *Cem. Concr. Res.*, **24**, No. 5, 813–829.

10.10   Scrivener, K.L. and Capmas, A. (1998) *Calcium aluminate cements*. 709–778.

10.11   Sharp, J.H., Lawrence, C.D. and Yang, R. (1999). Calcium sulfoaluminate cements — low energy cements, special cements or what? *Adv. Cem. Res.*, **11**, No. 1, 3–13.

10.12   Uchikawa, H., Hanehara, S. and Sawaki, D. (1997). Effect of electrostatic and steric repulsive force of organic admixture on the dispersion of cement particles in fresh cement paste. *10th ICCC Gothenburg, 1997*. Paper 3iii001.

10.13   Wang, S-D. and Scrivener, K.L. (1995). Hydration products of alkali-activated slag cement. *Cem. Concr. Res.*, **25**, No. 3, 561–571.

10.14   Zhang, L., Su, M. and Wang, Y. (1999). Development of the use of sulfo- and ferroaluminate cements in China. *Adv. Cem. Res.*, **11**, No. 1, 15–21.

Appendix

A1   Everett, D.H. (1967). (Flood, E.A. (ed.)). Adsorption hysteresis. *Solid gas interface*. Edward Arnold, London, **2**, 1055–1113.

A2   Gregg, S.J. and Sing, K.S.W. (1982). *Adsorption, surface area and porosity*. 2nd edn, Academic Press, London.

Index

accelerated curing 88, 125
accelerators 197, 198
acceptance limit values 80–81
activation energy
    in hydration 110, 111, 119, 125
    in pyroprocessing 45–47
admixtures 85, 93, 197–200
adsorption isotherms 143, 157
    pore size distribution from 142–144,
        209–210
adsorption,surface area from 75, 136,
    137, 138
aeration 86, 121
AFm phases 119, 120, 123, 126, 131, 136,
    180, 189
AFt phases 117, 120, 121, 122, 131, 133,
    136, 180, 189
aggregate 1, 2
    reaction with alkali 166–168
    see also concrete
air entrainment 200
alite 11
    formation 47, 56
    grindability 71
    hydration 123–124, 180
    NMR spectrum 106
    optical microscopy 55–57
    X-ray diffraction 59
    see also tricalcium silicate
akali activated aluminosilcate
    cements 204
alkali activated slag cements 186–187,
    203–204
alkali-silica reaction (ASR) 166–167, 185
    in blended cements 194–195
alkali sulfates 14, 48, 49, 58, 121
alumina ratio 5, 23
anhydrite 14, 19, 73, 100, 205, 206
aphthitalite (glaserite) 12, 14

arcanite 14
argillaceous raw materials 17–18
ASTM
    standards/test methods 75, 79, 85, 86,
        88, 168–169
    classification
    of cements 6, 163, 165
    of fly ash 177
autoclave curing 88, 125, 147, 148
autogenous grinding 17
autogenous shrinkage 100, 129
Avrami-Erofe'ev equation 110, 125, 129

backscattered electron (BSE) imaging 63,
    64, 71, 133, 142, 170, 171
ball mill 28, 67, 68
Bangham equation 155
belite 11–12, 57
    conversion to alite 47, 56
    grindability 71
    hydration 115–116, 124, 180
    optical microscopy 57
    sulfoaluminate cements 206–208
    X-ray diffraction 59
    see also dicalcium silicate
BET method 75, 136–137
Blaine method 74
blastfurnace slag
    alkali-silica reaction affected by 194,
        195
    composition 173, 185–187
blastfurnace slag cement 164, 187–188
    hydration 188–190
bleeding 91
blended cements 172
    alkali-silica reaction in 194–195
    sulfate resistance 191–193
Bogue composition, method 8–9, 25
    compared with other methods 60

bottle hydration   158, 159
British Standards   5, 79, 163, 164
    see also Portland cement
brucite (magnesium hydroxide)   125
burnability of raw materials   24
    see also combinability

C-S-A system   6–7
C-S-A-F system   6,8
C-S-H
    composition/structure   103, 108–109
    as hydration product of
        blended cements   180,185,189
        dicalcium silicate   115–116
        Portland cement   121–122,
            131–136, 138, 154–155
        tricalcium silicate   103–109
calcareous raw materials   15–17
calcareous fly ash   179
calcite (calcium carbonate)   15
    decomposition   3, 42–45
calcium aluminate cement (CAC)   173,
    201–203
calcium aluminoferrite   13–14, 59
    hydration   119–120
see also ferrite phase
calcium hydroxide (hydrated lime)   3
    effect on dormant period   113
    pozzolanic reaction   175
    see also portlandite
calcium langbeinite   14, 58, 121
calcium sulfate hemihydrate   73, 117
    heat of hydration   91, 121
calcium sulfate plasters   1–2
calcium sulfoaluminate cements   162,
    204–208
    expansive/shrinkage
        compensated   205–206
calcium sulfoaluminate-belite
    cements   206–208
capillary condensation   156, 209
capillary porosity   139
    relationship to elastic moduli   149, 150
cement, definition   1
cement chemist's notation   6
cementitious systems   1–2
chalk   15, 17, 21
chemical analysis by selective
    dissolution   54
chemical shrinkage   100, 129
Ciment Fondu   201

clay minerals   17–18, 21, 159
    decomposition of   41, 43, 176
    dehydrated   183
    fly ash from   176
clinker
    colour   53, 57
    compounds   11–14
    compound composition   23
        calculation of potential   8–9
        determination   58, 60, 63
    melt
        properties   46,48
        quantity   23
    production   31–49
clinkering
    mechanism   46–49
    reducing conditions in   47, 49, 53
coal ash   19–22
combinability   24–28
compacting factor   85
compressive strength
    British Standard   80, 82
    cement paste   146–148, 159
    factors influencing   83–84, 101,
        146–148, 158, 159
    fly ash cements   180–181
    slag cements   190
    test methods   82–85
computer modelling   160–161
concrete
    compressive strength   84
    durability/cracking   3, 88, 92–93,
        126–127, 145, 169–170, 207–208
    transition zone   132–133, 144, 185
    workability   85-86
conduction calorimetry   91, 92, 95, 96,
    109, 117, 121, 125
consistence, standard   81
    see also water demand
controlled-fineness cement   165
creep   151–153
cristobalite   167, 194
crystal growth pressure   88–89, 127, 169,
    205
cyclones   32

Darcy's equation   145
D-drying   104, 137
defects in crystals   10, 114
delayed ettringite formation (DEF)
    126–128

density
    clinker phases   55, 57
    Portland cement   74
    silica fume   184
diatomaceous earths/loams   183
dicalcium silicate   11–12
    hydration   115–116
    see also belite
differential thermal analysis   45, 100
diffusion
    in clinker melts   46–47
    in hydration   110–111
disjoining pressure   156
dormant (induction) period   3, 95, 109
    effect of fly ash   179–180
    origin   112–115
dreierkette(n)   108, 109
dry process   31
    fuel consumption   33, 39
    preparation of kiln feed   34–35
    pyroprocessing   38–40
drying of hydration products   98–99, 104,
    137–138, 152
drying shrinkage hardened cement
    paste   153–158
durability see concrete
dusting of clinker   12

elastic moduli   149–151
electrical conductivity   102, 182
electrical double layer   199–200
electron microscopy   62–64
energy consumption
    kiln processes   33, 52
    milling/grinding   34, 67, 69, 71
enthalpy changes in pyroprocessing   43
ettringite   6, 72, 91, 117–119, 123, 124,
    180
    disruptive formation   126–128, 169
    see also AFt phases
European classification of cements   164,
    172, 173, 184
European standards   6, 79, 80, 90,
    163–164
    fly ash   177
    blastfurnace slag   187
evaporable water   139
expansion   3, 87–88, 126, 166, 169,
    194–195
expansive cements   205–206

false set   86
    factors affecting   73, 121
    mortar test   86–87
Feldman-Sereda model   138, 154, 155,
    158
Feret's law   83
ferrite phase   13–14, 57, 59, 63
    hydration   98, 119–120, 124–125
fillers   72
filter press   33
fineness of cement   73–78
flash set   86
flexural strength   161
flint   16–17
flow properties, mortar/concrete   85, 86
fly ash
    alkali silica reaction affected by   194,
        195
    composition   176–179
fly ash cement, hydration   179–182
formulae, simplified oxide   6
free lime   7, 58
fuel consumption see energy consumption

gel   103, 136
    see also hardened cement paste
gel porosity   139, 181
gel space ratio   146, 161
geopolymer cements   204
Gibbs adsorption isotherm   154–155
grindability   28–29, 71
grinding aid   72
ground granulated blastfurnace slag
    185–187, 190
    see also blastfurnace slag cements
groundmass   132
grout   3
gypsum   1–2, 4, 19, 53, 72–73, 163, 201,
    205

Hadley grains   133
hardened cement paste
    compressive strength   146–148
    creep   152
    drying shrinkage   153–158
    elastic properties   149–151
    microstructure   131–136
    modelling   160–161
    permeability   144–146
    pore size distribution   141–144,
        209–210

porosity 139–141
solid-solid bonds in 147, 154, 157, 158–160
sulfate attack 168–171
surface area 136–138
hardening of cements 1
heat balance 51–52
heat of hydration 89–92
    effect of blastfurnace slag 190
    effect of fly ash 180
    values for Portland cement constituents 91
    variation with time 95, 96
heat recovery 40, 49
heat of solution 90
heat treatment *see* pyroprocessing
helium pyknometry 141, 142
high alumina cement (HAC) *see* calcium aluminate cement
high pressure grinding rolls 69–70
hindered adsorption 156, 157
Horomill 70–71
hydration 1
    at elevated temperature 125–128
    at surface mechanism 111, 112
    blastfurnace slag cement 188–190
    calcium aluminate cements 201–203
    calcium aluminoferrite 119–120
    dicalcium silicate 115–116
    effect of water/cement ratio 84, 133
    fly ash cements 179–182
    mechanisms 110, 111–112
    methods of monitoring 97–102
    Portland cement 120–125
    products, drying of 98–99, 104, 137–138
    through-solution mechanism 110, 111, 112, 119, 120, 133
    tricalcium aluminate 116–119
    tricalcium silicate 102, 103–115
    see also heat of hydration
hydraulic cements 1
hydraulic lime 4
hydraulicity 11, 101
hydrogarnet 189
hydrotalcite 125, 189
hysteresis 143, 151, 157

induction period 3, 112, 114 *see also* dormant period
initial set 80, 81

ink bottle pores 137, 142, 156
inner hydration products 112, 122, 134, 135, 136
intercalation 156, 157
interlayer hydration 137, 141
interstitial phases, optical microscopy 57

jennite 109

kaolinite 18, 43, 165
Kelvin equation 156, 209
kiln feed
    calculation of composition 20–23
    preparation 31–35
Kingery equation 46
Klein's compound 204
Knudsen equations 129–130

Langavant calorimeter 90
Laplace equation 155
Le Chatelier test 87–88
Lea & Nurse method 73, 74
Lea & Parker formula 20
Lepol (semi-dry) process 39–40
lime 3, 7
lime combination factor(LCF) 5
lime saturation factor (LSF) 5, 8, 20–21
    effect on combinability temperature 24
limestone
    as minor additional constituent 72
    as raw material 15–17
limit values 80
low alkali cement 165, 166–168
low heat cements 91, 165, 166, 180, 190

macro defect free cement 161–162
magnesia (magnesium oxide) 14, 58, 88, 187
    see also periclase
manufacturer's autocontrol 80
melilite 186, 188
mercury intrusion porosimetry (MIP) 141, 142, 145
merwinite 186, 188
microscopy *see* elecron; optical
microsilica 184
milling of cement 67–73
minor additional constituent 72, 63
minor phases (in clinker) 14, 58
mortar 3

mortar strength tests   82
Mössbauer spectroscopy   160

natural pozzolanas   182–184
nitrogen adsorption and surface area   75,
    136, 137, 138
nitrogen oxides (NO$_x$)   50
nodule formation   39–40, 46, 53
nomenclature/notation   6
non-evaporable water   98
nuclear magnetic resonance (NMR)
    spectroscopy   105
    application to
        alkali-silica reaction   167
        hydration   106–107, 111, 123, 202
    chemical shift   106–107

oilwell cements   200–201
opaline silica   166, 194
optical microscopy   54
    alite   55–57
    belite   57
    correlation with other methods   60,
        61–62
    interstitial phases   57
    minor phases   58
    quantitative   58
ordinary Portland cement   5–6, 73, 84
osmosis   114, 166
Ostwald ripening   127
outer hydration products   122, 134,
    135–136

particle size distribution
    determination   76
    effect on compressive strength   77–78
    fly ash   176
    Portland cement   76
        effect of classifier efficiency   68
        effect of mill type   77
        effect on combinability   24
paste   1, 3
    standard consistence   81
periclase   14, 58, 88
permeability
    concrete, factors affecting   92–93, 185
    hardened cement paste   144–146
permeabilty methods for surface area
    determination   74–75
pH changes in a cement paste   102,
    179–180, 189

plaster set   86
plasticisers   199–200
polymorphism   9–14
    see also clinker compounds
pore size distribution
    determination using
        adsorption isotherms   142–144,
            209–210
        BSE imaging   142
        mercury porosimetry   141–142
    hardened cement paste   141–144
porosity of hardened cement paste
    139–141
    and drying shrinkage   153–154
    and strength   146–148, 158, 159
    effect of fly ash   181–182
Portland cements   1
    ASTM classification   6, 163, 165
    BS12   5–6, 79–81, 163–165
    composition   5–9, 81
    controlled fineness   165
    European classification   163–164
    hydration of   120–125
    low alkali   165, 166, 168
    low heat   91, 165, 166
    ordinary   5–6, 73, 84, 165
    quality testing   81–91
    rapid hardening   73, 84, 123, 164
    sulfate resisting   25, 165, 172, 201
    white   5, 17, 164–165
portlandite (calcium hydroxide)   2,
    132–133, 144, 145
    reaction with
        fly ash   175, 180
        silica fume   185
pozzolanic materials   4, 173–185
    fly ash   176–182
    natural pozzolanas   182–184
    silica fume   184-185
pozzolanic reaction   174–176
precalciner   38, 39, 45
preheater/preheating   38, 41
process control   49–50
proportioning   8, 20–23
pulverised fuel ash (pfa)   173, 176
    see also fly ash
pyknometry   141, 142
pyrophyllite   18, 43
pyroprocessing   31
    enthalpy changes   43
    manufacturing processes   36–41

mechanisms involved   41–49
thermal efficiency   51–52

quality tests   79-93
quantitative X-ray diffraction analysis
   (QXDA)   60–62, 97, 98, 123, 180
quartz   167, 174

rapid hardening cement   73, 164, 165
   hydration   123
   strength development   84
raw feed/mix/meal   15, 31
   multicomponent   19
raw materials   15–20
   physical properties   28–29
   proportioning of   20–23
   reactivity   23–28
refractories   40–41
reinforced concrete, durability   3, 92,
   144–145
retarders   197, 198, 201
retrogression (in strength)   126, 201
rheology of mortar/concrete   85, 86
Rietveld method   61–62
roller (vertical spindle) mill   34–35, 69
roll press (high pressure grinding
   roll)   69–70
Rosin-Rammler equation   76
Rosin-Rammler-Sperling-Bennett (RRSB)
   diagram   76–77, 78
rotary kiln   31, 36, 37
rutile   59, 60

scanning electron microscopy (SEM)   62,
   64, 132, 133, 206
selective dissolution, chemical analysis
   by   54
semi-dry (Lepol) process   31, 34, 36,
   39–40, 52
semi-wet process   31, 32–34, 36, 38, 52
separator/classifier   35
set retarder/regulator   4, 19, 72, 201
setting of cements   1
setting time   81
   effect of fly ash   179
shale   17, 18
   burnt   183
shrinkage
   chemical   100

drying   153–158
shrinkage compensated cements   206
shrinking sphere model   44, 45
silica fume   184–185
   effect on alkali-silica reaction   195
silica ratio   5
   effect on combinability   24
sintering in the presence of a liquid
   see clinkering
slag,alkali activated   204
slag cements   187–190
slump test   85
solid solution in clinker compounds   9–14
solid-solid bonds   147, 154, 157, 158–160
soundness   87–89
standard strength   83
standard specification   79–80
steam curing   88, 125, 126
stiffening
   cement paste   81
   mortar   86, 87
   oilwell grout   200
strätlingite   189
strength see compressive strength, flexural
   strength
stress-strain plots   149–150
sulfate attack
   Portland cement   168–171
   blended cements   191–193
sulfate resisting Portland cement   25, 165,
   172
   oilwell cements based on   201
sulfates
   in clinker   14, 20
   maximum SO_3 in cements   88, 165,
   172
sulfoaluminate-belite cements   206–208
sulfoaluminates   6, 72, 88, 117
   see also calcium sulfoaluminate
      cements; ettringite
superplasticisers   199, 200
supersulfated cement   204
surface area
   cement   69, 71, 73
      determination   73, 74–75
      hydration affected by   109, 110
   hardened cement paste   136–138, 160
   silica fume   184
   strength,affected by   84
swelling clays   18, 32, 137, 157, 159
syngenite   201

thaumasite 170
thermogravimetry 100, 101
tobermorite 108, 109, 126, 138
    gel 108
transition zone *see* concrete
transmission electron microscopy
    (TEM) 62, 64, 103, 125, 134–136
tricalcium aluminate 13, 57, 59
    hydration 116–119, 124
tricalcium silicate 11
    hydration 102, 103–115, 124
    NMR spectrum 106
    strength of slag cements, effect
        on 190, 191
    see also alite
trimethylsilylation (TMS) 104–105, 107
tube mill 32, 67, 70
tuff 183

unsound cement 87

Van't Hoff equation 42
Vebe test 85
Vicat needle 81
viscosity of clinker melts 48
volcanic ash 182–183
    see also pozzolanic materials

Wagner method 75
wash mill 32
water
    enthalpy of evaporation 43
    non-evaporable 98
water adsorption
    determination of surface area 136,
        137, 138
    isotherms 143, 157
water/cement ratio

critical 140
effect on
    permeability 145, 146
    strength development 84
    stress-strain 149, 150
    surface area 137, 138
water demand 70, 77, 179
    see also workability
water reducers 197, 199–200
wet process 31
    fuel consumption 33
    preparation of kiln feed 32–34
    pyroprocessing 36, 37–38
    thermal efficiency 52
white cement 5, 17, 164–165
winning 15
workability 85–87
    effect of blastfurnace slag 190
    effect of fly ash 179
see also water demand

X-ray diffraction (XRD) 18, 58–62
    blastfurnace slag 186
    fly ash 178–179
    quantitative analysis (QXDA) 60–62
        hydration studied by 97, 98, 99,
            101, 123, 124
X-ray fluorescence spectrometry 53, 188
X-ray microanalysis 45, 63–64, 71
    application 127, 128, 133
xerogel 136, 160 *see also* gel
xonotlite 126

Young's modulus 149, 150

zeolites 183
zeta potential 199–200
zur Strassen equation 51